Human Origins

The Emergence & Evolution of the Human Species

"The first step in improving the human condition is taking an honest approach to understanding the true nature of reality."
– Don Hainesworth
July 11, 2013

DON HAINESWORTH

Gotham Books

30 N Gould St.

Ste. 20820, Sheridan, WY 82801

https://gothambooksinc.com/

Phone: 1 (307) 464-7800

Published by Gotham Books (July 16, 2022)

ISBN: 978-1-956349-91-7 P

ISBN: 978-1-956349-92-4 E

ISBN: 978-1-956349-93-1 H

Table of Contents

CHAPTER	TITLE	PAGE #
....	Preface	….. 04
....	Introduction	….. 05
....	Creation Myths	….. 07
1	Life from a Biological Perspective	….. 13
2	Origins of Life	….. 31
3	Evidence of Early Life	….. 46
4	Life in Extreme Conditions	….. 53
5	Evolution by Natural Selection	….. 70
6	Genetics	….. 81
7	DNA and Genes	….. 94
8	The Vertebrate Animal Evolution	….. 106
9	The Biology and Evolution of Becoming	….. 112
10	Human Evolution	….. 165
11	Neanderthal and Cro-Magnon	….. 192
12	Homo Sapiens (Modern Humans)	….. 270
13	Migrations of Modern Man and Diversity	….. 283
....	Appendix I	….. 339
....	Glossary	….. 345
....	Image Credits	….. 374
....	References	….. 387

PREFACE

*O*ur common ancestry with the great apes comes from fossil records. Our hominid heritage on the tree of life started on the continent of Africa. Around 18 million years ago this great land mass was separated or disconnected from the other land masses due to platonic tectonics floating around the Indian Ocean. It collided with Eurasia (now modern-day Europe)which eventually gave rise to the first African exodus. The apes that left at or around that time ended up in Southeast Asia, which evolved into gibbons and orangutans. And the ones that stayed on in Africa, evolved into gorillas, chimpanzees, and humans.

With respect to humans or Homo Sapiens, the root "homo" means "man" and the root "sapien" means "being. "Therefore, human beings. Human beings are bipedal primates belonging to the mammalian species of "Homosapien" in the family Hominidae (the great apes). About 1.5 to 2 million years ago there was Homo-habilis, the earliest form of man. He was the first 'great ape'. And about 500,000 years ago lived Homo Erectus (he walked on 2 legs). Then came the Neanderthal who lived about 100,000- 30,000 years ago. They were very primitive and animalistic, but very social. They lived in groups, wore clothing, used fire, and made basic hunting tools. Next came Cro-Magnum, who lived roughly, 40,000-10,000 years ago. Their cranial features were elongated to allow for a larger brain, they made more evolved tools by grinding rocks and were the developers of religion. The last and only surviving Hominid is a modern man who came into existence roughly 60,000 years ago out of Africa.

This book takes us through the very beginnings of life from the single-cellular organism to the multi-cellular all the way to the emergence and evolution of humankind to the present day.

INTRODUCTION
The Keys to Discovery

*W*e know that DNA is the—prime-mover inside nearly every cell in our body, built from a long string of nucleotides called T, G, C, A. And it's the order of these nucleotides that helps determine who we are and that tells us about our ancient family history.

But the long-sought-after answer to the riddle was how it all gives rise to becoming who we are. The unveiled mystery lies in how your DNA is packaged inside your cells. When we look at a section of DNA with its particular order of As, Cs, Gs, and Ts, it's a little bit like looking from a page in a book. Each page is lined-up alongside a lot of other pages, one right after the other in a row. Together this long string of pages composes a chapter in the story of who you are.

This chapter is what is referred to as a chromosome. A chromosome is a package of DNA that gives you a set of traits that make you who you are. Inside your cells, there are exactly 23 pairs of these chromosomes. Hidden on these chromosomes are little clues that can tell us exactly who you're related to going back thousands of generations, all the way back to the days of our earliest species.

To explain how this works in nature when you have children, your DNA is copied and passed on to them. And that's why children tend to look like their parents. All of those billions of As, Cs, Gs, and Ts are painstakingly duplicated in every generation.

Of course, the coping isn't perfect. Occasionally there are mistakes, like typos. These typos are called mutations. They are extremely rare. An average of just one hundred typos out of the billions of As, Cs, Gs, and Ts get passed on to offspring.

Most of the time, these mutations are harmless. In fact, they may serve a very useful purpose. They make each generation just a little different genetically, from the one before. Over many generations, these little mutations lead to the evolution of all sorts of useful things. Our opposable thumb, our large brains, etc. But most of these mutations have absolutely no effect on us.

They are simply ancestral baggage (or what is referred to as – junk DNA). And here's the most important part of the story. These little genetic mutations are passed on in the DNA of every descendant that follows.

What that means is that if you and someone else have the same mutations, you share an ancestor. The person in the past who first had that type in his or her DNA and passed it on to the two of you creates a marker of decent. And that's exactly what we call these typos, markers. These markers allow us to create a family tree for everyone alive today.

Everyone walking around on planet Earth will fall onto a branch of the Human family tree. And by examining our individual markers, we can know which branch of that tree we are on.

CREATION MYTHS

There are numerous myths concerning the creation of humanity and of the Universe. The following gives just a few excerpts from various accounts of how man and the Universe were created.

Hindu - This universe existed in the shape of darkness, unperceived, destitute of distinctive marks, unattainable by reasoning, unknowable, wholly immersed, as it were, in deep sleep. Then the Divine Self-existent, himself indiscernible but making all this, the great elements, and the rest, discernible, appeared with irresistible power, dispelling the darkness. He who can be perceived by the internal organ alone, who is subtle, indiscernible, and eternal, who contains all created beings and is inconceivable, shone forth of his own will. He, desiring to produce beings of many kinds from his own body, first with a thought created the waters, and placed his seed in them. That seed became a golden egg, in brilliancy equal to the Sun; in that egg, he himself was born as Brahma, the progenitor of the whole world… The Divine One resided in that egg for a whole year, then he himself by his thought divided it into two halves; and out of those two halves, he formed heaven and earth, between them the middle sphere, the eight points of the horizon, and the eternal abode of the waters. From himself, he also drew forth the mind, which is both real and unreal, likewise from the mind ego, which possesses the function of self-consciousness and is lordly. Moreover, the great one, the soul, and all products are affected by the three qualities, and, in their order, the five organs which perceive the objects of sensation. But, joining minute particles even of those six, which possess measureless power, with particles of himself, he created all beings.

Aztec - Coatlique was first impregnated by an obsidian knife and gave birth to Coyolxanuhqui, goddess of the moon, and to a group of male offspring, who became the stars. Then one-day Coatlique found a ball of feathers, which she tucked into her bosom. When she looked for it later, it was gone, at which time she realized that she was again pregnant. Her children, the moon, and the stars did not believe her story. Ashamed of their mother, they resolved to kill her. A goddess could only give birth once, to the original litter of divinity and no more. During the time that they were plotting her demise, Coatlique gave birth to the fiery god of war, Huitzilopochtli. With the help of a fire serpent, he destroyed his brothers and sister, murdering them in a rage. He beheaded Coyolxauhqui and threw her body into a deep gorge in a mountain, where it lies dismembered forever. The natural Cosmos of the Indians was born of catastrophe. The heavens literally crumbled to pieces. The Earth mother fell and was fertilized, while her children were torn apart by fratricide and then scattered and disjointed throughout the Universe. The mother of the Aztec creation story was called Coatlique (the Lady of the Skirt of Snakes). She was created in the image of the unknown, decorated with skulls, snakes, and lacerated hands. There are no cracks in her body, and she is a perfect monolith (a totality of intensity and self-containment, yet her features were square and decapitated).

Comanche - One day the Great Spirit collected swirls of dust from four directions in order to create the Comanche people. These people formed from the Earth had the strength of mighty storms. Unfortunately, a shape-shifting demon was also created and began to torment the people. The Great Spirit cast the demon into a bottomless pit. To seek revenge the demon took refuge in the fangs and stingers of poisonous creatures and continues to harm people every chance it gets.

An ancient Chinese creation myth - This myth dates to 600 B.C., Phan Ku the Giant Creator emerged from an egg and proceeded to create the world by using a chisel to carve valleys and mountains from the landscape. As our birth required the death of our creator, we were to be cursed with sorrow forever after.

Maya creation myth-- In the beginning was only Tepeu and Gucumatz. These two sat together and thought, and whatever they thought came into being. They thought Earth, and there it was. They thought mountains, and so there were. They thought of trees, sky, and animals. Each came into being. Because none of these creatures could praise them, they formed more advanced beings of clay. Because the clay beings fell apart when wet, they made beings out of wood; however, the wooden beings caused trouble on the Earth. The Gods sent a great flood to wipe out these beings so that they could start over. With the help of Mountain Lion, Coyote, Parrot, and Crow they fashioned four new beings. These four beings performed well and are the ancestors of the Quiché. After the Mik'Maq world was created and after the animals, birds, and plants were placed on the surface, Gisoolg caused a bolt of lightning to hit the surface of Ootsitgamoo. This bolt of lightning caused the formation of an image of a human body shape out of the sand. It was Glooscap who was first shaped out of the basic elements of the Mik'Maq world: sand.

Quiche Maya -The first men to be created and formed were called the Sorcerer of Fatal Laughter, the Sorcerer of Night, Unkempt, and the Black Sorcerer . . . They were endowed with intelligence, they succeeded in knowing all that there is in the world. When they looked, instantly they saw all that is around them and they contemplated, in turn, the arc of heaven and the round face of the Earth . . . [Then the Creator said]: —They know all ... what shall we do with them now? Let their sight reach only to that which is near; let them see only a little of the face of the earth! . . . Are they not by nature simple creatures of our making? Must they also be gods? - The Popol Vuh of the Quiche Maya.

Scandinavian -Odin is the All-Father. He is the oldest and most powerful of the Gods. Through the ages, he has ruled all things. He created heaven and earth, and he made man and gave him a soul. But even the All-Father was not the very first. In the beginning, there was no Earth, no sea, no sky. Only the emptiness of Ginnungagap, waiting to be filled. In the south, the fiery realm of Muspell came into being, and in the north, the icy realm of Niflheim. Fire and ice played across the emptiness. And in the center of nothingness, the air grew mild. Where the warm air from Muspell met the cold air from Niflheim, the ice began to thaw. As it dripped, it shaped itself into the form of a sleeping giant. His name was Ymir, and he was evil. As Ymir slept, he began to sweat. There grew beneath his left arm a male and a female, and from his legs, another male was created. These were the first frost giants, all of whom are descended from Ymir. Then the ice melt formed a cow, named Audhumla. Four rivers of milk flowed from her and fed Ymir. Audhumla nourished herself by licking the salty blocks of ice all around. By the end of her first day, she had uncovered the hair of a head. By the end of her second day, the whole head was exposed, and by the end of the third day there was a complete man, His name was Buri, and he was strong and handsome. Buri had a son named Bor, who married Bestla, the daughter of one of the frost giants. Bor and Bestla had three sons: Odin, Vili, and Ve.

Mesopotamian/Babylonian Creation Myth-When on high the heaven had not been named. The firm ground below had not been called by name. When primordial Apsu, their begetter. And Mummu- Tiamat, she who bore them all, Their waters mingled as a single body, No reed hut had sprung forth, no marshland had appeared, None of the gods had been brought into being, And none bore a name, and no destinies determined -- Then it was that the gods were formed in the midst of heaven. Lahmu and Lahamu were brought forth, by the name they were called.

The Enuma Elish is a Babylonian or Mesopotamian myth of creation recounting the struggle between cosmic order and chaos. It is basically a myth of the cycle of seasons. It is named after its opening words and was recited on the fourth day of the ancient Babylonian New Year's festival. The basic story exists in various forms in the area. This version is written in Akkadian, an old Babylonian dialect, and features Marduk, the patron deity of the city of Babylon. A similar earlier version in ancient Sumerian has Anu, Enil, and Ninurta as the heroes, suggesting that this version was adapted to justify the religious practices in the cult of Marduk in Babylon.

This version was written sometime in the 12th century BC in cuneiform on seven clay tablets. They were found in the middle 19th century in the ruins of the palace of Ashurbanipal in Nineveh. George Smith first published these texts in 1876 as The Chaldean Genesis. Because of the many parallels with the Genesis account, some historians concluded that the Genesis account was simply a rewriting of the Babylonian story.

Genesis (Bible –Old Testament)1:1
In the beginning God created the heavens and the earth. 1:2Now the earth was formless and empty. Darkness was on the surface of the deep. God's Spirit was hovering over the surface of the waters. 1:3God said, —Let there be light, and there was light. 1:4 God saw the light and saw that it was good. God divided the light from the darkness.1:5God called the light Day, and the darkness he called Night. There was evening and there was morning, one day.

1:6God said, —Let there be an expanse in the midst of the waters and let it divide the waters from the waters. 1:7God made the expanse and divided the waters which were under the expanse from the waters which were above the expanse, and it was so. 1:8God called the expanse sky. There was evening and there was morning, a second day.

1:9God said, —Let the waters under the sky be gathered together to one place and let the dry land appear,‖ and it was so. 1:10God called the dry land Earth, and the gathering together of the waters He called Seas. God saw that it was good. 1:13There was evening and there was morning, the third day.

1:14God said, —Let there be lights in the expanse of the sky to divide the day from the night; and let them be for signs, and for seasons, and for days and years; 1:15and let them be for lights in the expanse of the sky to give light on the Earth, and it was so.

1:16God made the two great lights: the greater light to rule the day, and the lesser light to rule the night. He also made the stars.

1:17God set them in the expanse of the sky to give light to the earth, 1:18and to rule over the day and over the night, and to divide the light from the darkness. God saw that it was good. 1:19 There was evening and there was morning, the fourth day.

Genesis (Modern Version) **1** At the beginning when God created* the heavens and the earth, **2** the earth was a formless void and darkness covered the face of the deep, while a wind from God* swept over the face of the waters. **3** Then God said, Let there be light, and there was light. **4** And God saw that the light was good, and God separated the light from the darkness. **5** God called the light Day, and the darkness he called Night. And there was evening and there was morning, the first day.

6 And God said, Let there be a dome in the midst of the waters, and let it separate the waters from the waters.' **7** So God made the dome and separated the waters that were under the dome from the waters that were above the dome. And it was so. **8** God called the dome Sky. And there was evening and there was morning, the second day. **9** And God said, Let the waters under the sky be gathered together into one place, and let the dry land appear.' And it was so. **10** God called the dry land Earth, and the waters that were gathered together he called Seas. And God saw that it was good.

11 Then God said, Let the earth put forth vegetation: plants yielding seed, and fruit trees of every kind on earth that bear fruit with the seed in it.' And it was so. **12** The earth brought forth vegetation: plants yielding seeds of every kind, and trees of every kind bearing fruit with the seed in it. And God saw that it was good. **13** And there was evening and there was morning, the third day. **14** And God said, Let there be lights in the dome of the sky to separate the day from the night and let them be for signs and for seasons and for days and years, **15** and let them be lights in the dome of the sky to give light upon the earth.' And it was so. **16** God made the two great lights—the greater light to rule the day and the lesser light to rule the night— and the stars. **17** God set them in the dome of the sky to give light upon the earth, **18** to rule over the day and over the night, and to separate the light from the darkness. And God saw that it was good. **19** And there was evening and there was morning, the fourth day.

20 And God said, Let the waters bring forth swarms of living creatures, and let birds fly above the earth across the dome of the sky.' **21** So God created the great sea monsters and every living creature that moves, of every kind, with which the waters swarm, and every winged bird of every kind. And God saw that it was good. **22** God blessed them, saying, Be fruitful and multiply and fill the waters in the seas, and let birds multiply on the earth.' **23** And there was evening and there was morning, the fifth day. **24** And God said, Let the earth bring forth living creatures of every kind: cattle and creeping things and wild animals of the earth of every kind.' And it was so. **25** God made the wild animals of the earth of every kind, and the cattle of every kind, and everything that creeps upon the ground of every kind. And God saw that it was good.

26 Then God said, Let us make humankind* in our image, according to our likeness; and let them have dominion over the fish of the sea, and over the birds of the air, and over the cattle, and over all the wild animals of the earth,* and over every creeping thing that creeps upon the earth.' **27** So God created humankind* in his image, in the image of God he created them;*male and female he created them. **28** God blessed them, and God said to them, Be fruitful and multiply, and fill the earth and subdue it; and have dominion over the fish of the sea and over the birds of the air and over every living thing that moves upon the earth.' **29** God said, See, I have given you every plant yielding seed that is upon the face of all the earth, and every tree with seed in its fruit; you shall have them for food. **30** And to every beast of the earth, and to every bird of the air, and to everything that creeps on the earth, everything that has the breath of life, I have given every green plant for food.' And it was so.

31 God saw everything that he had made, and indeed, it was very good. And there was evening and there was morning, the sixth day.

CHAPTER 1
LIFE FROM A BIOLOGICAL PERSPECTIVE
1.1 A Scientific Definition

What is *life* ? Well, that may seem like a very strange question to ask. And for most of us, it may seem that the answer is obvious. If you look around at the sorts of things that you're familiar with on a day-to-day basis, like some of the creatures shown in this slide, it seems obvious when something is alive. But it turns out that the answer to this question is not quite as obvious as it seems. We might wonder why we would even want to ask that question. Why would it have been, and why would it be a concern to biologists? Asked the question, what is life? Well, if we're looking for life elsewhere, for example, on another planet, we need to be able to say what life is. What is it that we're actually looking for? And to do that we need some sort of definition of life.

If we're trying to understand the origin of life on the earth, we need to know what we're talking about. What are we actually looking for in the rocks? What sort of evidence in the early fossil record, for example, the life on the earth we are looking for. When is something living and when is something not? So, biologists are searching for life in the universe but before they do that, they need to know what is life and they need to be able to tell people what it is they are looking for and that's why this question is so important. It also turns out of course the question, of what is life has huge social implications, like in vitro fertilization, for example. We've seen all sorts of arguments on the media about when is a human embryo alive. Well, we're not going to discuss those questions in this course but the question of what is life. And when does something become alive, has social implications as well as implications for biology? So, let's have a look at some of the characteristics that you and I might think define living things. Well, if you look at pet dogs, for instance, they seem to be very complex, and they exhibit complex behaviors. We might look at that behavior, we might look at the behavior of our pet dog, the way it greets us, and we might say that kind of complexity is indicative of life. Non-life doesn't exhibit the same sort of complexity as life does. It also seems to grow. Again, if you've ever had a pet dog, it seems fairly obvious that it starts off as a puppy, and it grows into an adult dog.

And if you look around at most of life on Earth, it seems to grow. This seems to be a characteristic of living things. And you might think, well, that's a characteristic of life. And we might list that as another type of feature that we associate with living things. Life also replicates. That may be fairly obvious. If it didn't, it wouldn't persist for very long. And again, you can see this with dogs. They replicate, they produce puppies.

And this seems to be a characteristic of all life on Earth that many people are familiar with. Replication seems to be necessary for life to persist on our planet for many generations. Life force metabolizes, that means a simple term to eat, it needs a source of energy in order to be able to grow and reproduce and do some of the other characteristics that we've just talked about, so we might say that one characteristic of life is that it metabolizes, it eats food, it makes energy. Life also has a system for storing information. Okay, so life on earth, is the double helix of DNA, and that information storage system is necessary to pass on information from one generation to another. But also, to be able to program the cells in our bodies that allow them to grow and carry out the functions they need to do to be living in the first place. Life also seems to evolve by Darwinian evolution. Early dogs, for instance, probably look something like this wolf. And today, of course, dogs look very, very different. It's not just a process of natural selection. In the case of dogs, they have been artificially selected by human beings, artificially selected for particular characteristics.

But nevertheless, this is a type of selection. And it illustrates, very well, the process of evolution, by which, organisms can evolve and change, and in a natural environment, evolve and change to cope with changes in their environment and alterations in the natural habitat in which they live. The ability to adapt to change, and to evolve in the Darwinian sense also seems to be a characteristic of life. And indeed, this ability to evolve has even been used as a definition of life. Here is a definition by Gerald Joyce, a NASA scientist who described life as a self-sustaining chemical system capable of undergoing Darwinian evolution. There is just one example of a type of written definition: a simple definition of all life on the Earth that we might apply to life on other planetary bodies. But hang on a second. We need to be very careful. Because many of the characteristics that I've just described for life are things we also find in the non-biological world. This is a picture of a tornado. And as you can see it seems to exhibit some quite complex behaviors. Forms these twisting funnels of air which we say that because it's complex and it exhibits complex behaviors it is alive. Well, I think most of us would agree that a tornado is not alive. Many things in the non-biological world also grow. These salt crystals, just a few millimeters in size, can grow into these large hand-held crystals that you can see here. And yet this crystal is not alive, despite the fact, that it seems to be growing.

Replication seems to be a much more robust character of life, and yet other things replicate we don't think of being alive. Computer programs, for example, can replicate from one computer to another. Indeed, when you put a file on a memory stick, and you share it with your friends you're replicating a file from one computer to another. But you wouldn't say that that memory stick is alive in any way. And yet, the information on it is being replicated. And even non-living things seem to metabolize in a very broad way.

This fire, for example, is burning up trees. And as it's burning trees, it's releasing energy. In fact, in the same chemical reaction that occurs. In our own bodies as we burn out organic compounds of food with oxygen, we make energy. This is called aerobic respiration. In a forest fire, the trees are burnt in oxygen and make energy. The chemical reaction is exactly the same is just a little bit less controlled than the chemical reactions inside our own bodies. But in a very broad way could say this forest fire is metabolizing and yet most of us probably wouldn't say That a forest fire is alive. There are other complications as well. For example, many things that look biological seem to lack characteristics that we expect from life. For example, viruses cannot replicate on their own. They need a host cell to divide. Does that mean that viruses are not alive?

Well, some people think that's the case, that you cannot include viruses in a definition of life. Other people think it's absurd, that something like a virus that has such a profound impact on our bodies, cannot be considered alive, particularly when it contains nucleic acids, such as DNA. Many other living things cannot replicate on their own. Here is a rather strange example. A rabbit cannot replicate on its own. It

15

needs another rabbit. Does that mean that a rabbit on its own is not life? And only when it's with another rabbit, with which it can replicate, does it become life?

Well, I think all of us would consider that to be rather an absurd question. I think most people would consider their pet rabbit, even when it's on its own, to be alive. But this shows you some of the complications that we get into when we try to define life using particular characteristics. Evolution seems to be a very robust way to define life.

And we've seen that there's even a definition of life-based on the ability of organisms to undergo Darwinian evolution. But in recent years, we've seen computer programs developed that seem to be able to evolve and change over time. Would we say these computer programs are alive? Well maybe, at some point in the future, when computers become much more complex, that might be an argument worth having on whether computers are alive or not. But these early programs that seem to evolve and change inside a computer, we would not say are really, truly alive. So even the characteristic of evolution may not necessarily be a defining characteristic of life, and this brings us to a possibility, a problem, that the definition of life. The possibility is that life is just a definition, a human definition. There is no real physical or chemical characteristic about life that we can use to find what it is. It's just A human definition that's useful to us, and this is what philosophers refer to as a non-natural kind. Some things in the world are referred to as natural kinds. For example, the element gold is a natural kind, and what this means is it has. This is a very distinct definition. We can define its density. We can define its atomic weight. We can define its melting point. And so on and so forth. If someone says to you, what is gold, you can tell that person exactly what it is, and you can list the physical properties of that element that define gold. And we call such a material a natural kind. A non-natural kind Is an object that is defined by human definition, and a good definition of this is a chair, we might define a chair as an object that you sit on, and then what happens if we sit on a table does that mean then a table is also a chair, and so on and so forth, and you can see how we can get in all sorts of circular arguments? Defining what is a chair, what is a table, and what is the difference between these objects is an example of a non-natural kind and ultimately it doesn't matter what you define as a chair or a table, it's just a human definition.

And you can define it however you like, as long as you're consistent and other people understand the definition that you were using. Maybe life is the same. Maybe life is just a word. It defines a part of chemistry that we like to think of as encapsulating our ideas of life and we can draw that line between life and non-life wherever we want. As long as we all agree on what we're talking about, and that definition's agreed upon among scientists. It doesn't really matter where you define the line. It's really just an empty human definition, but nevertheless, a definition that's important for biology. So, we should consider the possibility that there is no sharp definition of life. It's a human definition useful to us. Let's have a look at some of the history of attempts to understand the nature of life. And what is life? And how we might define it? Well, surprisingly, like a lot of things, the ancient Greeks gave thought to the definition of life, and what constitutes life. Their school of thinking is sometimes called materialism. And they had an idea that life has a soul. And the soul is made of atoms of fire. And this follows from the thoughts of Empedocles, one of the early Greek philosophers, who asserted that all things are made of earth, water, fire, and air. And because life seems very lively, it's made up of atoms of fire. And these early thoughts about the definition of life show us something very important, right from the beginning the very early definitions, people thought there was something very special about life, something that separates us from the non-living.

For the Greeks, it was having a soul and being made of a particular type of atomic material. As time developed and people start to develop, we moved into the European enlightenment. And during the enlightenment, scientists had a similar sort of idea to the ancient Greeks. It could vitalize the idea that life contains some vital force and that when this vital force is added into some non-biological it becomes biological. It is the same sort of idea, the idea that life has some special type of characteristic. That makes it categorically different from non-living. And this led to all sorts of ideas around the idea of spontaneous generation. Spontaneous generosity is the concept that when a vital force is added to non-living matter it can become living. So, meat, for example, when it gains this vital force turns into maggots, so as a result, dust, and wheat when it gains some vital force turn into mice. And these were some of the ideas that dominated thinking about the origin and nature of life.

In the early years of the seventeenth century, early thinkers in the enlightenment even came up with some very, very bizarre experiments. And to us, these experiments seem utterly ridiculous, but they were taken quite seriously in the earlier stages of scientific study. This is an experiment, an example of a documented experiment from the seventeenth century. And the idea is that you take a jar. And you put into that jar some wheat husks and some old underwear. And you seal the jar, and you leave it for 21 days. And during those 21 days, a vital force will move into the jar. And it will transform the underwear and the wheat husks into mice. When you open that jar 21 days later, you will find mice. Now, you really have to question the powers of observation of those early scientists that added wheat husks into a jar and didn't recognize that there were baby mice in there, but do remember, this was the very early years of the Enlightenment, and the scientific method was not fully developed, and people didn't understand the idea of controlled experiments.

These are the sort of things that people wrote down and very rapidly became folklore and accepted. Knowledge – the knowledge that vital forces could transform non-living into living matter. There were many attempts to disprove spontaneous generation let's just look at two of them. They are rather interesting because they show how the scientific method began to develop and with it more definitive ideas about the possible nature of life. This is a rather ingenious experiment that was developed by Fransesco Redi, an Italian doctor, and he had a hunch that maggots might not be produced by spontaneous generation but might have something to do with flies buzzing around meat. So, he did this experiment where he put pieces of meat on the surface of the slab, and in two of the experiments, he. He covered them.

In one of the experiments, he covered it with a metal lid which prevented anything from getting in or out and landing on the surface of the meat. In another Experiment, he had a piece of meat on a slab, and he covered it with gauze. The gauze would allow through a vital force if it really did exist, but it would stop flies from getting to the meat. And in the final experiment, he just left the meat out in the open. So that it could exchange anything with the outside world. What did he observe? Well, unsurprisingly, a few days later what he observed was that on the meat that was completely covered in the metal mitt, there were no maggots.

In the meat covered in gauze, there were also no maggots. But in the meat that was left open, there were maggots. And the interpretation of this experiment is that flies were required to somehow interact with the meat, lay their eggs as it turns out, and form maggots. But in the case where the meat was covered in the gauze, if there was a vital force out there that took hold of the meat and turned it into maggots, we would expect that meat to also give rise to maggots. But it didn't. And this was one experiment that brought to an end the idea of spontaneous generation and created a more empirical basis for understanding the nature of life. This is the second experiment, and possibly the most elegant experiment, that finally brought an end to the idea of spontaneous generation. It's an experiment that was designed by famous microbiologist Louis Pasteur. In 1859, he won a prize from the French Academy of Sciences. He would challenge scientists to come up with an experiment that would end the idea of spontaneous generation. He developed these swan-necked flasks. In which he put broth. And in one of these flasks, he boiled the broth, to kill off all the life. And this swan neck that you can see on the flask, prevented any type of microbes or other life from getting into the flask. And once he had boiled the broth and he left these flasks for many days, nothing happened to it. It remained completely as it had been after he boiled it. But in one of the flasks, he broke off the swan neck, so opening up the flask to the outside world. And unsurprisingly after a few days, this broth became turbid, microorganisms it turned out had landed in the flask, we're growing in the broth. And it caused it to be filled with life. And this was another experiment that showed that if you take a breath, it doesn't just spontaneously give rise to life. It has to be accessed, it has to have microbes land in the breath and be able to grow within. This experiment, with one of the most elegant experiments, showed that microorganisms were also responsible for causing life to emerge in different materials and that they had to move around and be able to access the material in order to create life. And this was the first experiment that really ended spontaneous generation. So, we have this history, and we've got these ideas about the definition of life. Where are we today?

Well, most biologists today I think accept the idea that life might just be a working definition. And whatever types of definitions we do take for life, we can probably find exceptions to the rule. But most biologists also accept that we use working definitions to define life. They are useful. We can use them to search for life on other planets, as long as we all agree on what those basic ideas about life might be and what some of the characteristics of life that we're looking for on other planets. As long as we keep an open mind when we're searching for it elsewhere maybe there is something that's completely different. Life as we don't know it. So as scientists when we explore other planets, we keep an open mind to the possibility of types of alien life that may not fit our definition of life. But these definitions do have to allow us to search for life elsewhere. It allows us to study the structure of most of the life that we're familiar with, in a consistent and rational way.

So, so far, we've learned that attempts to define life have been made since the ancients. And that in fact, it's very difficult to define life accurately. It may even be impossible to define life accurately if it's just a human definition. But despite that, we can develop working definitions of life that are useful in biology. And we can use those definitions to define what it is that we're searching for on other planets. But we need to be cautious, and we need to have an open mind as scientists to the possibility there might be types of life that don't quite fit our own definition of life.

1.2 Structure of Life - Building Blocks

We've looked at the definition of life, now let's look at the basic structure of life. How it's made up, and what are its fundamental units? Well, this is what you and I see on a large scale. These are just some of the organisms that we're all familiar with. How are they actually built? What are the building blocks? Well, of course, unsurprisingly, life is made up of elements, the elements of the Periodic Table. And when you look at a periodic table, it looks very complex. There are well over 100 elements there. How can all those elements come together to form life? And how can we get to the bottom of understanding which of these elements are important? Well, it turns out that the fundamental structure of life is really just based on a few of those elements, a very few of them. And this is a list of them here. They are in fact Carbon, Hydrogen, Nitrogen, Oxygen, Phosphorus, and Sulfur. Sometimes it's referred to as the word Chenops. And these are the six elements that make up most of the compounds from which life is constructed. Throughout the elemental level, life is actually quite simple. Of course, it goes without saying that other elements in the periodic table are used by life. For example, there's iron in our red blood cells which is why we need iron in our food and there's calcium in our bones which form the phosphate structures that are part of all skeletons. That's why of course we drink milk to get our calcium. So other elements are in life, but when we look at all life on Earth and we try to find the common elements that bind all those life forms together it turns out there are very, very few of them. There are just six elements that make our task of understanding life a lot simpler. So how do we take these elements and put them together to start to fill, and form the building blocks of life? Well, these elements or atoms come together to form molecules and I've shown here an example of one particular molecule called Glycine which is an amino acid, one of the building blocks of proteins. And Glycine is made up of some very simple elements. There are two carbon atoms there. There is also nitrogen, some hydrogen, and oxygen, so you can see that even Glycine conforms to the idea of CHNOPS. It's constructed of some of those six basic elements. And you'll notice a couple of things about this molecule.

First of all, it's quite simple. And that's true of most of the molecules, which are the building blocks of life. They're very simple structures made up of a few atoms. And the other thing you'll notice is that carbon is the backbone of this particular molecule. In fact, there are two carbon atoms in glycine. And other atoms attached to those carbon atoms. Atoms. We think of carbon as the backbone of this molecule. And this is why we refer to life on earth as being carbon-based life. Because the molecules from which you and I are constructed have as their backbone carbon, it's the common element that binds all these other elements together. So, we're carbon-based life. Now, these molecules are very simple, and you might think well then how can you form a much more complex organism. Well, the molecules come together to make more complex molecules. I've just shown you glycine, an amino acid. If we take these amino acids and we link them together in a long chain, we end up with a protein, and a protein is essentially a long chain of amino acids. And you can think of stringing them together like beads on a string. Once they are attached together, they form these long chains, and those chains can start to do complex things like bind to other chains and fold up in particular shapes and form the complex diversity of complex molecules that make up life. Proteins, for example, make up much of your body. And here are some examples of foods that contain proteins, essential for gaining those amino acids that your body uses to construct more complex proteins, and other types of proteins, necessary for your body to function. And on the right day, you can see some of the protein supplements that people eat in order to get hold of these amino acids that are so essential to building blocks for constructing the proteins in our cells. Of course, life needs other types of molecules as well. For example, sugars, and this is an example of some sugars. And you can see on the left-hand side there are some healthy sugars and of course, we're all familiar with some unhealthy sugars as well. And on the right-hand side there you can see a structure. Of a typical sugar. It has this ring-like structure. If you look at it, it conforms again to what we've said earlier. You can see fine carbon atoms that form this ring around oxygen and then various other atoms attached to those that ring of carbon atoms. Again, this is Carbon-based life.

It's a molecule that's constructed around the backbone of carbon with other atoms attached to it and this particular sugar could be strung together a bit like those amino acids in a protein to create long complex chains of sugars and that's how we end up. These complex carbohydrates are what we find in our foods. Sugars are another type of building block that can be put together with other types of molecules to construct the complexity of life. And its component molecules and cells. Some of these molecules we eat, as we've just seen in proteins and sugars, and some of these molecules we make in our body or are made by life inside the cell. This is just a rather nice example of some of the complexity that gets created inside the cell. On the left-hand side there, you can see this ring structure which is called Adenine. And Adenine can be transformed by adding a ribose sugar structure that you can see in the center there, to make this compound called adenosine. We can take adenosine and we can add phosphorus atoms to it and oxygen atoms, these phosphate molecules as they're called, phosphate groups. And we end up with this molecule called adenosine triphosphate. If we take adenosine triphosphate and we string it together with a variety of other molecules, it forms a component of deoxyribonucleic acid DNA, the information storage molecule of life. And so, this is just a nice example of how the body goes from a very simple molecule, like adenine, made up of a few carbon atoms, some nitrogen atoms, and some hydrogen and strings it together, with other molecules to create something as complex as the genetic code that allows for the transfer of material from one generation to another. All these reactions, these biosynthesis reactions as they are called, occur within the cells. In all organisms on the earth, this is the way in which life goes from basic elements on the periodic table to very complex molecules. It is essentially like Lego, constructing complex cellular structures from very simple component parts or building blocks. Now we can't do these reactions. We can't make these molecules in a dry state. These molecules have to be moved around. They have to be added to one another. Chemical reactions have to occur whereby these molecules can grow more complex molecules. We have to do that in a liquid as a liquid allows these molecules to move around. And on earth, all life carries out these reactions. In the liquid which is water.

And that's why we say that life is also water-based. It's not just carbon-based, it's also, water-based. Water is the solvent, if you like, the solvent in which these biochemical reactions can occur. So, the two basic things that life needs are those six elements: Carbon, hydrogen, nitrogen, oxygen, phosphorus, and sulfur, to build the building blocks of life. And then it needs water in which these reactions can occur. And we'll look in another lecture about whether we really need these sorts of elements, or whether alternative life forms might use different pathways. But certainly, all life on earth has this common architecture of these elements and the use of liquid water as a solvent So those are the basic building blocks of life, what have we learned in this lecture We've learned that life, is, of course, made of elements. But it's made of very basic ones, including carbon, hydrogen, nitrogen, oxygen, phosphorus, and sulfur.

We've learned that life uses other elements as well, depending on how it's built. An example of this is calcium used in bones. These elements come together to form molecules including amino acids and sugars, the building blocks of life. These molecules come themselves, come together to form even larger structures such as proteins and complex sugars, and other molecules such as DNA, Deoxyribonucleic Acid. We've learned that life is a little bit like Lego. It's built up from small components into much more complicated ones and we've learned that life also needs water in order to carry out these biochemical reactions, these reactions to build up from elements to molecular building blocks, and finally, the complex structures that we think about like life.

1.3 *Structure of Life- Cells*

We've looked at the structure of life, the basic building blocks from which it's constructed. Now let's go a little higher up the hierarchy and look at the structures that are formed from these molecules. So here are our friends again. These are molecules that we are familiar with on a large scale. What happens if we take them, and look at them under the microscope? Well, one of the first things that we notice is that they're made up of cells. And this is an image of some skin cells, essential packages of reactions, and molecules that form cellular structures. Cellular structures can be something of a few microns size, a few thousandths of a millimeter in diameter. And cells are quite complex, if you break them apart, you'll find many functions within a cell that are responsible for allowing life to do its various chemical reactions. But in essence, there are three major features of cells that we're going to look at here that are important from a biological point of view in the sense that these are the things that define how these cells function, and how they make up life. First of all, there's a membrane, a membrane to enclose all those chemical reactions. If you didn't have a membrane, you would simply dissipate into the environment. All of those macro-molecules, or those large molecules that we saw earlier, would just disperse into the environment. So, cells, essentially, have a membrane to enclose the biochemical reactions. They also need an information storage system, we've seen that that's deoxyribonucleic acid, DNA, in order to encode the instructions that the organism needs to keep up its cells, grow cells, and reproduce. And that information storage system also transmits information. From one generation to another. So, cells must have within them, a coding structure and information storage system. And they must also have a system for gaining energy. Where are they going to get the energy to do all those chemical reactions in building up those, building blocks that we saw earlier? And how are those cells going to reproduce? Well, they need the energy to do that. So, within each cell, there must be an apparatus to make energy. So, a membrane, an information storage system, and an energy-producing system are basic to all cells on Earth. If you look at a mammal cell, it's very, very complex. It includes all sorts of organelles.

These other small structures inside the cell do other things, important for communication for example between different mammal cells, and for encoding other instructions that allow the cell to move around and do other complex patterns of behavior. But even if you look at a complex cell-like this from a mammal, you'll find that it still has those three Basic units. There are three basic ingredients that it needs in order to be able to function. A membrane encloses everything, the information storage system and DNA, and a method for gathering energy. Let's have a look at those three structures and think about them in more detail, and why they're important for making life work. First of all, let's look at the membrane. The cell membrane is made up of compounds called lipids. We saw earlier how these are building blocks, structures, and complex molecules that makeup life. Lipids are another type of complex molecule. They're very ingenious because some of these lipids have, as you can see here, a head that's attracted to water, and a long tail that doesn't like water. And that tail wants to get away from water. If you take these molecules and you put them in water and you mix them up, those tails spontaneously are attracted towards each other because they're trying to get away from the water, but the heads are attracted towards the water. And, as you can see here, what you get formed is essentially what's called a bi-lipid membrane layer. It's a layer of these molecules that spontaneously forms a membrane in water. And that layer of molecules can form micelles.

Spherical structures, or near-spherical structures that can enclose material inside. So, a cell membrane is made up of these phospholipids, these lipid molecules that spontaneously arrange in liquid. Automatically, they arrange together to form these, these micelles, these membranous structures that can enclose things. Within them, once we form the membrane in this way, we then have to form the information story system and this is made up of DNA, Deoxyribonucleic acid, let's have a little, little, look at this molecule. And see how it works, see how it functions in its role both as an information storage system but also in transmitting generation, in transmitting information from one generation to another. DNA has a backbone as you can see in this double helix and between this double helix are what we call base pairs. And these base pairs are made up of four different molecules.

Adenine is what is called A. Thymine that we'll call T. Cytosine that we'll call C. And guanineis called G. These are the four letters of an alphabet if you like, a code that instructs the cell on how to make an articular Molecule. For example, in one cell, if we were to string those particular molecules together, it might read something like AATCGCAG and that might mean building this particular molecule responsible for eye color. Another piece of DNA might have a different sequence of these letters, GGCAGAT, and maybe that means to construct a molecule that's responsible for helping to build bones, and so on and so forth. By stringing these four letters together in different combinations we can create codes of long strings of the letter that tell the cell to do particular things. It's a really ingenious code. With particular codes, the discrete codes are called genes. And genes are a set of instructions that tell the cell to produce a particular protein or a particular molecule. So, DNA is a molecule in which these four smaller molecules, A T C, and G are strung together In a long code. And provide the information stored for that cell in order to be able to do its housekeeping functions and make new types of molecules. So, you might then wonder. Well, okay. That explains how DNA acts as an information storage system. We can see how this molecule can use these four letters to create long ingenious codes for producing particular molecules. But how does it replicate from one generation to another? Well, DNA has another very intriguing and ingenious trick up its sleeve. These four letters we've just spoken about A,T, & G, and C can only bind to each other in particular combinations. A can only bind to T and C can only bind to G. And they bind together to form these base pairs and you can see a diagram here of base pairs of a long string of these letters on the left-hand side and the complementary basis on the other side. So, where we've got a C. Attached to it, we have a G. And where there's a T, there's an A and vice versa. And so, throughout the center of DNA, there are these base pairs of T-A and G-C, bound together through the center of the molecule. Let's have a look at what happens if we pull that base, those base pairs apart, if we split the DNA double helix in two. Well, if we split the double helix in two, we end up with two single strands. And along one strand we've got that code that we've just talked about, and on the other strand, we've got the complementary code, those letters that previously bound to the other strand.

We've just said that G can only bind to C. So, if there's a G in that single strand of DNA, the cell knows that it has to bind a C back onto that G. And next to a T it has to bind an A, and so on and so forth. So, if we split the two single strands apart, we can immediately synthesize a complementary strand. On both of those new strands that are now single strands, and that way we can generate two new double helices, which you can see here depicted in this diagram. So, the DNA double helix is split in two, we've ended up with two single strands and then the DNA knows from that code, which Of those four letters, it must bind to the letters already there in the single strand. And once it's done that, it can re-synthesize two strands of DNA. And this is the way in which DNA replicates information from one generation to another by faithfully replicating information from one cell to another as it divides. And this is how cells divide. And here's a diagram of the bacterium dividing, and the DNA is splitting into, the double helix is splitting into single strands that are being used to re-synthesize double strands, and as a result, two cells can form from one. This is the basis of DNA replication and cell division. It's what allows our cells to grow and it's what allows organisms to grow from puppies to large dogs as we saw earlier in the course. So, we can see now very simply how we can enclose information in a membrane and make it sound, how we can use DNA a very ingenious molecule, to instruct the cell to produce new molecules and how it can divide. What about energy? All cells need the energy to grow and reproduce. To get that energy, to divide DNA in the first place all cells contain systems for gathering energy. And there are different types of energy systems. Here are just three of them, to give you some idea. You and I, we're called heterotrophs. And that means we get our carbon from organic carbon like steak, or other types of meat. Organic molecules are made of carbon. There are other types of organisms in the biosphere, which we'll come across throughout this course, that use different methods of getting energy. For example, the phototrophs like trees, gather their energy from sunlight. Photosynthetic Organisms are called phototrophs, which literally mean seating in sunlight. And then there are organisms called chemolithotrophs, which literally means chemical rock eater, chemical element eater. These are creatures that could get their energy by eating rocks.

And these are confined to the microorganisms, these are literally bacteria for example, that get their energy by consuming iron in rocks. And these turn out to be very interesting for biology. Because the chemolithotrophs live in very extreme environments, like volcanic environments, and deep in the crust of the earth, where there's no organic material to feed them, like you and I need, and there's no sunlight, they can't do photosynthesis. Instead, they live off the energy from rocks. These key melithotrones are the sorts of organisms we might look for deep in the crusts of other planetary bodies such as Mars. So, these are just three examples of energy gathering, photosynthesis from light, heterotrophy by etiorganic molecules including us, and chemolithotropes, micro-organisms that get their energy from rocks. This is a diagram of the energy-gathering apparatus in cells. We don't need to worry about this, but I wanted to show you it because you can see it's actually very, very complex. The biochemical machinery together with energy from the environment has an enormous number of molecules involved. But at the end of the day, if you strip away all that complex biochemistry, it's very, very simple. It's all about extracting energy from atoms and molecules in the environment, to provide the energy for cell replication and cell growth. You might think that getting energy from rocks, might be biochemically a complex thing for a cell to do, but at the end of the day, it's using some pretty simple materials, the elements within rocks, that you and I might find outside in our garden. So, life looks very diverse, and it looks very unrelated. And when you look at it on a large scale, particularly the sort of organisms that you can see with the naked eye such as these ones here, it may look like that diversity makes life very complex to understand and indeed life is very complex, but at its core, it's very, very simple. And what we've seen in this lecture Is that the cellular structures from which life, life is made. Cellular structures built up from those building blocks that we saw earlier lecture are actually common to all life on earth. Energy gathering, cellular membranes to enclose the biochemical reactions body of life and, and information storage system, DNA, used by life on earth to transmit, energy, to transmit information from one generation to another, And to encode instructions in the cell. What have we learned, then, in this lecture? We've learned that life is made of cells.

We've learned that cells have some basic features: a membrane, an information storage system, and the ability to gather energy, although depending upon the type of cell, they also have many other organelles and complex structures. These are the three basic building blocks of cells. We have learned all life on earth has these characteristics. In other words, all life on earth has some very simple. Systems from which the cells are made, and which have been the case since the origin of life on the Earth. (Cockell 2013)

CHAPTER 2
ORIGINS OF LIFE
2.1 The Building Blocks of Life

One of the unsolved puzzles of biology is how did life originate? Where did the building blocks for life come from? There is a central conundrum that biologists face in trying to address the question of how we go from basic molecules to self-replicating microorganisms. And I should say right from the beginning we don't know the answer to this problem, but many significant strides have been made by biologists. In understanding where those building blocks came from and how they might have been assembled into the earliest life forms. So, there are several possibilities of how life might have originated, and here are three of them. The first one is some supernatural or divine intervention. Life was created by some supernatural force.

Well, whether you believe that or not, it doesn't get off very far. It doesn't explain how the building blocks of life were created and how they might have assembled into the early living organisms on the Earth. So even if you were to believe this, it doesn't get away from the central biological problem of understanding those building blocks. Another response to how life originated might be it originated elsewhere and was transferred to the Earth, maybe on comets or in asteroid impacts. This is a possibility, we can't prove at the current time that life originated on the Earth, but it still doesn't get us away from the problem of how life originated in that distant location? Now of course there's a problem there because if it originated on a planet with very different conditions, then there had been conditions on some different world that were much more conducive to the origin of life. But without the ability to test that at the current time, without the ability to find other planets and prove that life originated there, we're left with our final possibility and that's that life originated on the earth. And it's that assumption that we use in trying to understand the origin of life, and at least constrain the sort of experiments that biologists do, in trying to understand how building blocks were assembled in early life and how they came together.

So, the first question we might ask as biologists is where did the building blocks of life come from in the first place? Recall from an earlier section that life is constructed from building blocks such as amino acids and sugars. One of the earliest experiments to look at where the building blocks of life might have come from was carried out by Harold Urey and Stanley Miller in 1952, and it's really become an iconic experiment. In the design of this experiment, they take a container with water and boil that water, and the water travels through a tube to another container that contains a simulated early Earth atmosphere. And that atmosphere has methane, hydrogen, carbon monoxide, and ammonia. And across that chamber, they have an electrical discharge that put energy into the gasses. And these chemical reactions occur as the gasses are electrified to this discharge and as the water passes through them. They carried out this experiment and after a few days and weeks, they found the production of a yellow substance almost a tarry substance in their experiment. And they removed that substance and analyzed it and they found that it contained amino acids as well as other complex chemical compounds. This experiment shows that you could potentially create the building blocks of life, amino acids, and the building blocks of proteins in simple experiments such as this. Well, nowadays, we know that the early Earth's atmosphere was slightly different from the atmosphere used by Urey and Miller. We think it was less reducing. In other words, it had fewer of the compounds, ammonia, hydrogen, and carbon monoxide. It may have been more oxidizing. More content of carbon dioxide in nitrogen. But nevertheless, this experiment was really a turning point in Biology. Because, for the first time, it allowed us to think about the creation of these, the formation of these building blocks of life in simple chemical reactions. It turns out that amino acids are really everywhere in the universe. We find them in meteorites that land on the surface of the earth. This is an image of the Murchison meteorite that landed on the earth in the late 1960s. And it was found to contain amino acids. Aspartic acid, glutamic acid, and many other types of amino acids that are used in biology are the building blocks of proteins. It turns out that these meteorites that also contain amino acids are not used by biology on Earth, such as non-biological amino acids.

It seems that life has selected a subset of the available amino acids to be found in the extraterrestrial environment. And since this early work, many other meteorites, like carbonaceous chondrites have also been found to have amino acids. This suggests that these are common constituents of extraterrestrial material. What does all this tell us? It tells us that the building blocks of life could have been produced on the Earth, endogenously in chemical reactions on the surface of the early Earth. And it also tells us that the building blocks of life could have arrived on the Earth in extraterrestrial materials in meteorites. So, we have two sources of the building blocks of life for early chemical reactions. Amino Acids are the building blocks of Proteins, and that's significant if you think that the early history of life began with Proteins building up with cellular materials and finally into life. Another question is where nucleic acids come from, such as DNA, Deoxyribonucleic Acid that we talked about in the earlier lectures, as the information storage system of life. Well, there's another type of Nucleic Acid called RNA, Ribonucleic acid it's very much like DNA.

Instead, the Thymine of that four-letter code in DNA is replaced by Uracil it has a four-letter code, Adenine, Guanine, Uracil, and Cytosine. These four letters come together in a chain a bit like DNA, forming these strands of RNA. RNA is a very interesting molecule because it seems to be a little bit more primitive than DNA. What's particularly interesting about RNA is it can self-assemble into molecules called ribozymes, and these ribozymes can carry out chemical reactions even self-replication. And one idea is that the early life on earth began from RNA an RNA world as it's been called. These RNA molecules would be floating around perhaps in pools of water on the early earth carrying out chemical reactions replicating and leading to early genetic systems, which would eventually transition to more complex chemical systems and eventually to early cells. One of the questions that biologists address is, which came first, protein world proteins constructed from these building blocks of amino acids that may have come from space, or endogenously produced in environments and earlier, or did life start in this RNA world produced from these early nuclear gasses. Well, it may well be that both came together. Perhaps proteins and RNA were necessary to construct the earliest cells and these reactions were occurring at the same time.

This is one of the questions that biologists currently seeking to address. What were the earliest molecules that led to the earliest forms of chemistry, that in turn led to early life forms, and in what sort of environments were these molecules produced? We saw that in an earlier lecture, we need the building blocks of life, amino acids to build proteins.

2.2 Nucleotides To Build Nucleic Acids

We also saw that life needs membranes in order to create cells. And one question is, how do you get membranes? Membranes would've been very important for early life on earth to enclose those early chemical systems preventing them from dissipating into the environment and simply becoming diluted in the early. Were early water bodies on the earth. How would these early chemistries have been enclosed in membranes? Well, experiments studying the production of early membranes have shown that you could produce membranes very, very easily. Scientists have taken materials from simulated interstellar ice, and even materials directly from meteorites.

These materials are added to water and found that membranes can spontaneously form from the lipids inside these materials. It turns out that meteorites contain lipids and these lipids when they're added to water. The tails of the lipids are attracted to each other. The heads of the lipids, which are hydrophilic, like water, are attracted to the water, and you get this spontaneous formation of vesicles in water, early membranes, which could have enclosed early chemical reactions on the earlier So in a very simple way we could think of some sort of scheme of meteorites coming in from space and other types of material, landing in the early oceans and dissipating in the water.

And perhaps chemical reactions on the early Earth as well. Forming lipids that spontaneously form membranes, vehicles in which chemical reactions could occur. Amino acids, formed from early chemical reactions on the Earth, and also brought in by meteorites, would have provided the building blocks for early proteins, and other chemical reactions could have provided the early building blocks for sugars and nucleic acids, RNA, and eventually DNA. And in these sorts of environments, These materials would have assembled to form early cells. Well of course you can tell by the way I say that it's very hand waving and very speculative. In truth, we really don't know how these membranes and these early building blocks came together to form a self-replicating cell. It's one thing to create a membrane and to use that membrane to enclose some basic chemical reactions. But going from that to an organism self-replicating looks something like bacteria or archaea that we're familiar with on today's Earth is another matter altogether. And it remains one of the great challenges in Biology. Understand that transition from the chemical world to the biological world. So, what do we learn in this lecture? We have learned that the

building blocks of life could be produced in simulated early earth conditions. We have also learned those building blocks could be found in extraterrestrial material. So, it could have been produced on the early earth.

They also could have been delivered to the surface of the early earth. In extraterrestrial materials, we've also learned that primitive membranes that can enclose chemical reactions necessary for the first cells to form can be easily made in the laboratory from simple molecules, even meteoritic materials, and simulated. Materials formed on interstellar ices. But we've also learned that the order of the emergence of the building blocks is not known. But what is important is that biology has now at least identified plausible sources of these early building blocks and some of the processes by which they may have come together and led to the first life forms on earth.

2.3 The Origins of Life - Location

We've looked at the building blocks of the origin of life and thought about some of the processes by which early life might have come to be, but another question is where did this occur? Where did these early reactions occur and where are plausible environments where the origin of life Might have happened. Well, in order to answer that question or think about answering it, we, first of all, have to understand what we need for the origin of life. Quite apart from those building blocks, there are three other things that are necessary in order to get an origin of life. First of all, we need an energy source.

We need the energy to do those chemical reactions, to create those building blocks, and eventually, to assemble those building blocks into more complex structures and eventually cells. We also need a means of concentrating molecules. In an absence of membranes, those early molecules that we need to assemble early membranes and other chemical reactions have to be able to be concentrated otherwise they'll just dissipate in the environment. They'll become diluted in the oceans.

So, we need environments where complex chemicals can begin to concentrate. And finally, we need an environment that's conducive to these complex molecules and their assembly once they've been made. It's no good having an environment for example that's so hot that when we assemble membranes they simply fall apart because of the intense heat. So, we need to find environments where the physical and chemical conditions might be conducive to these molecules once we've made them. One of the earliest speculations about the origin of life was made by Charles Darwin, who in 1871 sent a letter to his friend Joseph Hooker wrote the following, and it's worth reading it out, it's rather remarkable speculation. But if, and oh, what a big if, we could conceive in some warm little pond, with all sorts of ammonia and phosphoric sorts, light, heat, electricity, etcetera, present. That a protein compound is chemically formed, ready to undergo still more complex changes.

In the present day, such matter would be instantly devoured or absorbed, which would not be in the case before living creatures were formed. Darwin's warm little pond has become rather an iconic image of where the origin of life might have occurred. It's certainly a good idea.

But based on what we know about environments today on the earth can we come up with perhaps some more concrete speculations about what that warm little pond and what those environments might have looked Look like. Let's have a look at some of the environments where people think the origin of life could have occurred. One possible environment is deep-sea vents. These are vents at the bottom of the ocean were reducing fluids, and hot fluids by gushing up from the crust of the earth into the oceans. These fluids contain iron and sulfur, which is very interesting because in some of the energy transfer proteins in life we find chemical compounds that contain iron and sulfur. Some people have speculated that these are complexes, that may be a remnant of early life on the earth. Alkaline deep-sea vents are thought to be particularly favorable places. The formation of chemical compounds the source of energy in this environment comes from the hot fluids coming up from inside the crust.

The concentration of organic compounds could potentially occur in the chimneys around these vents where they're gushing up into the oceans, and in the rocks that are formed around the edges of these plumes of materials. That's being produced in the oceans. They are maybe conducive to the formation of complex compounds because the compounds could be dissipated into the oceans or collected around the vent where temperatures are cooler and where they wouldn't break apart so easily. So, we can see that many of the requirements for the production of chemical compounds needed in the origin of life might have been, might have been met in these deep-sea vents. Another possible location is impact craters. An asteroid or comet's impacts on the early Earth were much more common than impact events today and in these environments. There may have been places where early chemical compounds required for the origin of life could have formed.

When an asteroid or comet slams into the surface of the Earth several things happen, first of all, it heats up the rock providing a source of energy for chemical reactions. Secondly, that heat can create hydrothermal cells, in other words, circulating water through the impact craters. That circulated water could've provided environments for early chemical reactions to occur to produce complex compounds. And the water that collects inside that impact crater might be an almost literal interpretation of Darwin's warm little pond.

A body of water is heated by that, by the energy created that came from the impact, creating an environment where early chemical compounds could form inside the impact crater. As the impact crater cools down so conditions will become more conducive to complex chemical compounds to assemble and form inside the crater as temperatures became less extreme. So, asteroid craters are yet another possible location for the formation of early compounds and their assembly into early life forms. Yet another environment that people thought about are beaches. Beaches are interesting because of course as the tide comes and goes. Water flows over the surface of rocks. And as the tide goes back out again, the water inside the rocks begins to evaporate, concentrating chemical compounds inside those rocks. So perhaps rocks on the edges of beaches were also places where early chemical compounds could have concentrated. And early chemical reactions could have occurred to form the building blocks of life and ultimately more complex chemical compounds. A bit like the hydrothermal vents, yet another location could be a volcanic hot spring in the early continental land masses instead of in the deep oceans now on the continents, and the Advantages of these early volcanic hot springs would have been much like the deep-sea hydrothermal events.

The source of energy would've been hot fluids coming from volcanic regions in the crust, flowing onto the surface of the earth. Chemical compounds would have collected in these early volcanic hot springs and as some of these hot springs began to cool down and circulate into regions that are even cooler. There would have been conducive conditions for complex organic compounds and ultimately cellular material to have formed in these hot springs. Here's another very interesting idea for how the early building blocks of life might have formed in bubbles. And the idea is that volcanoes erupted onto the sea releasing gases enclosed in bubbles. And these gases might have included hydrogen. Methane forms in other types of reactive gases. The gases concentrated inside the bubbles then reacted to produce simple organic compounds. These bubbles rise to the surface of the ocean and burst to release their contents into the air. These simple chemical compounds, these simple organic compounds would then drift through the atmosphere and react even further, perhaps with ultraviolet radiation in the atmosphere from the sunlight, creating more complex organic compounds.

And eventually, these complex organic compounds will rain back into the ocean as raindrops and rejoin that process. Again, forming bubbles and being re-circulated to the surface of the ocean, and back out into the atmosphere. There's no direct evidence for this process as a way to form the early compounds required for life, but it's a very interesting alternative possibility for the production of early organic compounds. Of course, we've also seen that organized compounds could have been delivered from outer space in meteorites. So, there are other possibilities other than some of the environments we've looked at on the early Earth. Yet, another possibility is that all of these environments were reactors producing the early compounds required for life.

Many scientists get very focused on one particular environment. It's not surprising people do research on volcanic environments, or impact craters, or for example production of organics in early seawater. But it's quite possible every one of these environments is generating organic compounds, perhaps the whole of the early Earth was a giant prebiotic reactor producing chemical compounds in different environments that came together in one particular environment to produce early cells, an environment. That we don't know the identity yet, but early environments would have all been producing these early chemical reactions. Can we find any evidence in life for the environment in which it evolved? Is there anything about the biology of life that tells us where it might have evolved? What's very interesting is that some of the most primitive microorganisms at least some of the most.

Deep branching as we call them, micro-organisms that seem to be the most ancient, are heat-loving micro-organisms. Thermophiles are literally heat lovers. And these micro-organisms grow in volcanic vents, and also in deep-sea vents and they may be some indication that the earliest life forms on Earth, were actually heat-loving microbes that lived in hot environments. We have to be careful though because the early. Earth was bombarded by asteroid and comet impacts. And because of volcanic activity, it was much hotter than it is today. So it might be that these. Ancient microorganisms that love heat represent some sort of bottleneck in evolution, microorganisms that were capable of surviving hot conditions on the early Earth but not necessarily the earliest organisms that represent the earliest stages of life when it first evolved. But nevertheless, the fact that some of the most primitive organisms on the Earth seem to be heat-loving is rather intriguing. There are other more intriguing environments that are quite counter-intuitive, where early reactions for life on earth might have occurred, and one possibility is in ice sheets. Water has an interesting property that when you freeze it, it tends to exclude salts that collect in boundaries between ice grains. Here you can see an image of some ice crystals and between them, the boundaries where salts might collect, and in those boundary lines, we might find concentrated chemical compounds where earlier reactions could have occurred. Of course, at low temperatures, chemical reactions occur much more slowly than at higher temperatures but with many millions of years for these reactions to occur Maybe ice sheets on the early Earth were the places where early life could have evolved. Now, we have no evidence, of course, that early life evolved in ice sheets or the possibility that early chemical compounds would have formed in ice sheets. But it reminds us that we need to keep an open mind about the possibilities for the early origin of life. There may be some quiet, unusual environments that we think are not plausible locations for early chemistry but might turn out to be important as places for early chemical reactions. So, what have we learned in this chapter?

Well, hopefully, we've learned that numerous environments could have provided plausible locations for the production of the building blocks of life, and we've looked at some examples of those possible environments. We've learned that these environments need not be mutually exclusive and different molecules might have been produced in different places. Indeed, the whole of the early earth might have been a giant prebiotic reactor. And finally, we don't know how quickly this happened. Did the origin of life occur within a matter of days or weeks, or did it take hundreds of millions of years? This is one of the unsolved questions in the origin of life and a question that biologists still have to address.

2.4 The Origins of Life – Alternative Chemistries

We've looked at the building blocks of life and we've looked at some of the environments in which life might have evolved. But one thing we've assumed throughout this. Is that life is based on carbon, and it uses liquid water as a solvent. Let's revisit that assumption and think about whether there are other, plausible alternatives that have been considered by biologists. First of all, we said that life is based on carbon, and what that means is that the molecules of life have carbon as their backbones. Here are two examples of molecules.

Methane, which has a single carbon atom and four hydrogen atoms, is a simple compound produced by life but is still based on carbon. And on the right-hand side there, is the much more complex molecule of deoxyribonucleic acid, or DNA which is also carbon-based. The molecule is constructed around carbon atoms, with other atoms attached to it, such as hydrogen, nitrogen, phosphorus, and oxygen. These are carbon-based molecules. And all of our molecules are built up in the same way, using carbon. As the basic, building block of these molecules. What are the reasons why carbon is such a good atom-performing molecule? Well, one reason is it tends to form bonds with other atoms that are similar in energy.

For example, it can form bonds with hydrogen nitrogen oxygen and phosphorus. And the reason why it's important is that those bond energies between carbon and those other atoms are relatively similar. It means that carbon can move around those different atoms essentially exchanging them and forming different chemical compounds. It can break a bond with one atom and form a bond with another type of atom without needing much energy or without giving up much energy. And that creates a great versatility in the sort of chemical compounds that you can form. Another reason why carbon is so favored is that once it does form a bond with these other atoms, those chemical compounds are quite stable. So, the results of this versatility of carbon being able to bond with other types of atoms and the stability of the resulting bonds and molecules are that we can produce quite complex organic compounds such as this DNA molecule here. One alternative that has been discussed by biologists is silicon, silicon-based lifeforms have been, a favorable alternative for science fiction writers and also in films as well.

Here's rather interesting speculation from H.G. Wells in the novel 1894. And he says, one is startled towards fantastic imaginings. Visions of silicon-aluminum organisms or visions of silicon-aluminum men have been imagined. Wandering through an atmosphere of gaseous sulfur, let us say, by the shores of a sea of liquid iron some thousand degrees or so above the temperature of a blast furnace. But silicon has some major disadvantages. For example, it forms very stable Chemical compounds with oxygen. In fact, you can see these compounds by looking out your window and looking at rocks. Rocks are silicate, where essentially these are compounds where silicon is bound with oxygen to produce very stable silicate minerals. One of the best-known examples is caught. So, on any planet where there's oxygen lying around it will tend to react with silicon and form these silicate rocks.

These very stable compounds prevent silicon from engaging in other chemical compounds that might be of interest to biologists as plausible building blocks for life. Of course, we could imagine planets with very, very low concentrations of oxygen would free up silicon to get involved with other types of reactions, but it seems that silicon has too great a propensity to form reactions with tiny amounts of oxygen to form these stable silicate compounds for it to be useful. In the origin of life, another problem with some silicon compounds is they're very reactive. For example, here are two simple molecules, one of which is a carbon-based molecule, methane? We know that we can ignite methane, we use it in our gas ovens. And on the right is the corresponding silicon compound silane, which spontaneously ignites at room temperature. It's a very reactive compound. Now that doesn't mean to say that life could not use these compounds on a planet where temperatures are much cooler. Chemical reactions might occur much more slowly and deal with some of the problems with these highly reactive silicon compounds. Nevertheless, this might be one disadvantage of silicon, some silicon compounds on a planet like earth. What about elements other than silicon? Well, if we look through the periodic table at other possible elements that we might use as building blocks for life, unfortunately, they don't fare much better than silicon. For example, gases like helium and neon are too inactive to be the basis of chemical building blocks for life.

Oxygen, nitrogen, boron, and other types of atoms have a limited number of bonds to other atoms and therefore are not going to form the complexity of the compound that we associate with carbon. And elements like magnesium, calcium, potassium, and sodium tend to form ionic bonds is difficult to form bonds with many elements in the same way that we can do with carbon compounds. And so, as we look across the periodic table and we look at the advantages and disadvantages of different elements, what we find is that carbon really comes out on top as the best element to form stable complexes and highly diverse molecules.

Life also needs a solvent in which to carry out its chemical reactions so for life on earth that's liquid water. Water has many advantages as a solvent for life e.g., it readily dissolves many chemicals making it possible to carry out chemical reactions. Another intriguing advantage of water, some people have said it's necessary for life. Because when it freezes and becomes ice, it floats rather than sinks. And, as a result, life, like these fish, can live under a lake that has a frozen surface. It's speculated that if life is the solvent that salt when it became solid, then lakes and ponds would simply freeze through completely and prevent life from inhabiting the surface of a planet. Well, of course, we can argue about whether that's true or not, but it is an intriguing property of water that is very beneficial for life living in cold environments. Water also has a very wide temperature range. You find liquid water in the polar regions, and we find liquid water in boiling hot volcanic springs this wide temperature range allows life to carry out chemical reactions in many diverse environments on the Earth. What about other alternatives? Well, one alternative that's being favored by science fiction writers and even scientists are ammonia. Ammonia is interesting because at low temperatures it also readily dissolves many chemicals, and it could potentially allow for ammonia-based bonds between chemicals similar for example, amino acids some people have proposed Primitive protein chains made from ammonia-based chemical reactions. Problems with ammonia are that it has a very narrow temperature range. It's the only liquid from -78 to -34. But perhaps that's not a problem on a planet where there are large surfaces in which these temperature ranges are met. And also, it sinks when it freezes. Meaning that in bodies of liquid ammonia when they become very cold and then become solid, they would freeze through completely.

And how do we know that there aren't life forms that adapt to surviving for long periods of time in frozen ammonia and thawing out when the ammonia is melted again. Of course, these are speculations that we can't address. There are other types of solvents that are being proposed for life as well, such as hydrogen fluoride. Hydrogen fluoride when it's a liquid has a very wide temperature range, from -83 to +20 C and it also dissolves a wide range of substances. The problem with hydrogen fluoride is that fluorine is quite rare in the universe. It's 100,000 times less abundant than oxygen, necessary to form water. And it's rather aggressive at destroying organic compounds. So, on the face of it, liquid Hydrogen fluoride also doesn't look like a particularly good solvent for life. It's interesting to observe that our apparent toy narrow view of life is consistent with what we see in the Universe. Some people have said that our idea of Carbon-based life forms, using liquid water, is just very narrow. Earth-centered view of life and that on other planets we would expect to find life with entirely novel biochemistries, almost unimaginable to us but, in fact, when we go to other planets in our solar system, we don't observe unusual life forms. For example, on the surface of Venus, there are 464 degrees centigrade.

We don't observe strange silicon-based life forms. Using some unknown solvents, we observe what appears to be a lifeless surface. And so, it seems that our prejudices about carbon-based life using liquid water may not be so far from the truth, as a plausible view, of how life might be constructed through the rest of the universe. But a question that does face biologists, is - are there other bio spaces? Our bio space, based on life using carbon as a building block for its molecules and liquid water as a solvent Might just be one by space. Are there spaces using, for example, silicon-based lifeforms in liquid ammonia? Are there spaces that use carbon-based molecules in liquid ammonia? A future challenge in biology is to really determine whether these speculations about other types of chemical compounds, silicon-based compounds, and other types of solvents, like ammonia and hydrogen fluoride are really plausible and whether they really do occur on other planets or whether they are just fanciful ideas of science fiction. So, what have we learned in this lecture? Well, we've learned that carbon Is the most versatile element for building molecules. At least as far as we can understand based on our current knowledge of chemistry. We've also learned water is the most versatile and useful substance for doing chemical reactions. We've learned that there are other alternatives such as silicon as a basis for molecules and ammonia instead of water. We also learned the fact that silicon ammonia and other compounds seem less favorable than carbon-water as a basis for life. This doesn't rule them out. The carbon-water bio-space seems to be the most plausible or likely to be the most common architecture for other life but as biologists, we should keep an open mind. There may well be other planets out there. Perhaps rare in the, in which these alternative chemistries are being experimented with, as the basis of life. (Cockell 2013)

CHAPTER 3

EVIDENCE OF EARLY LIFE
3.1 Early Earth Conditions

We've seen that the conditions on the early earth became clement for life rather early on, but what is the evidence for early life on Earth? The only direct evidence we have is from fossils and chemical signatures preserved in the geological record. And this evidence may be proof that life was established on the earth 3.5 billion years ago, possibly as early as 3.8 billion years ago. It's thought that if the evidence does suggest life by 3.8 billion years, then it's likely to have arisen sometime before that because that early evidence probably was probably not preserved in the fossil record. This would suggest that life was present soon after the formation of the crostinations and that the end of that period of late, heavy bombardment, that period of the intense asteroid and comet impact, on the surface of early Earth.

Now before we look at some of that evidence, it's worth reminding ourselves that the evidence is subject to intense scientific controversy and debate. In fact, the evidence, of early life on Earth, is one of the most hotly debated areas of biology. There are several lines of evidence that have been developed to suggest the presence of life on early Earth. One line of evidence of early life on Earth is features called stromatolite.

Stromatolites are laminated mounds that form today in shallow marine water. You can find them in Shark Bay in Australia, for example. They're built up by successive accumulations of sediments in microbes, and also rocks such as calcium carbonate. So as the microbes form, sediments collect on top of them, and new layers of microbes form, and eventually, you end up with these macroscopically visible mounds of microbial activity. The oldest stromatolites have been found in 3.46-billion-year-old silica rocks, in Western Australia. One of the reasons why they're so subject to the controversy is because these mounds as you can understand look like features that could form by non-biological processes. For example, sediments that lay down in shallow rim environments or rivers for example. They can also form these wavy textures that look at least on the large scale, a little bit like stromatolites.

As a result, further lines of evidence have been in search of micro- fossils. Fossils of individual microorganisms in these stromatolites or other types of rocks could be evidence of life. The 3.46-billion-year-old stromatolites in Western Australia contain these filamentous structures, which are thought to have resembled modern cyanobacteria. They're composed of kerosene, which is the alteration product of heated and pressurized organic matter, and you can see some examples here of purported filaments in these. These microfossils have also been subject to intense debate, particularly because non-biological processes can also form fundamental structures. We don't really know the context of these ancient rocks. So, this is another line of evidence that biologists seek in order to demonstrate early life on earth such as fossils of individual micro-organisms. There are also indirect ways to show the presence of life. And one way is through chemical fossils. For example, carbon, which we met earlier in this course, and is the backbone of most molecules used by life, comes in different forms or isotopes. For example, there's carbon 12, which is the most common type of carbon in the environment, and carbon 13. Carbon 13 has 7 neutrons, carbon 12 has 6 neutrons. In other words, they have a slightly different atomic mass. Life will preferentially use this lighter isotope carbon 12. So, wherever there's been life, the carbon in those molecules within that life will tend to be made up of carbon 12, rather than the much rarer carbon 13, and this preferential uptake of this isotope carbon 12 can be used as evidence for the presence of life. And you can see some calcium

carbonate shells here, in which these early signatures can be preserved. These are much more recent fossils. But similar sorts of organisms taking up carbon through photosynthesis, depositing that carbon either in minerals or in organic matter can also be used to find chemical signatures on early earth. This so-called isotope fractionation this preferential uptake of light, lighter isotopes compared to heavier isotopes by life is seen in carbonate rocks. It was far back as 3.5 billion years ago and slightly more controversially earlier, as well. So what is the problem with all of this evidence, these three lines of evidence? Microscopic features such as stromatolites, microfossils of individual microorganisms, and indirect evidence such as chemical signatures. Well, first of all, these are rocks very heavily metamorphosed. In other words, they've been altered by heat and pressure since their formation those many billions of years ago. This makes it much more difficult to interpret the evidence. The geological settings are usually uncertain or heavily disputed. We don't know exactly what the conditions were like in the regions where those early rocks were formed, and so were they really conducive for life? That's very important to determine, to find out whether the fossils or chemical signatures that we see are really plausible evidence for life. Many of the features that we observe, can be produced, by non-biological processes. The way texture of Stromatolite can be produced in sediments, without biology. Micro-fossils can be formed with similar types of structures with non-biological processes; filamentors; non-biological structures that just look like biology.

Chemical signatures, called isotopic fractionation, can also be caused by non-biological processes. But scientists think that when this evidence is taken together, it does provide some compelling evidence for life on the early Earth. So, what have we learned? we've learned that it's thought that life was established on the Earth at least 3.5 billion years ago. The main evidence of this life comes from microscopic features such as stromotolites, microfossils of creatures An independent chemical alteration such as the fraction of carbon isotopes. Many people dispute this evidence on the grounds that similar features could be the result of non-biological processes, but it does show that the search for life on early Earth and the evidence for life on early Earth is an ongoing challenge of biology, and this work will reveal whether the early life on earth was established 3.5 billion years ago or earlier, and what the nature of this life might have been.

3.2 The Tree of Life

One challenge of biology is to understand the relationship between particular organisms. And how that early life on Earth gave rise to the multiplicity of life we see on the Earth today. And this is the job of phylogeneticists. What we're going to talk about here is the tree of life. The tree of life is essentially a depiction of the relationship between organisms that evolutionary relationships and how they branch from one another. This is an example of a tree of life, and you can see it's made up of many groups of organisms. In fact, there are three groups or domains of organisms. On the left you can see bacteria such as cyanobacteria, these are photosynthetic cyanobacteria that produce oxygen. And on the far right-hand side, part of that tree of life includes animals which include, of course, ourselves, as well as fungi and plants.

How do we create this tree and how do biologists know the relationship between these different organisms? Well, phylogenetics is the field of study that seeks to understand the evolutionary relationship between groups of organisms. It's based on comparisons of molecules within those organisms such as the genetic storage information, DNA, or differences between their shapes. The tree of life helps address several interesting questions in biology. Such as, which organisms appeared first? Who lived on the early Earth? What was, what was the nature of those early organisms that inhabited the early Earth? And what is the history of life on Earth? How did those early organisms on Earth eventually branch out into the organisms we're familiar with today?

Let's have a look again at this tree of life and some of its features. Here's a more detailed tree of life. And you can see that it starts with a common branch. And this is the common ancestor, the common ancestor of life that gave rise to all life on Earth today. It's the job of biologists to try and find out what that common ancestor was. The common ancestor gave rise To 3 main domains of life. These 3 domains are bacteria, which include many of the bacteria that are living in your soil in your garden, also bacteria that cause disease. And then another domain called archaea. Aarchaea is made up of some of the organisms that live in the most

Extreme environments on the Earth, such as microbes that live in hot springs in the deep oceans or in the volcanic springs in volcanic areas.

And then on the right of that diagram, you can see the eukaryotes and eucaryota. And eucaryotes include all the multi-cellular life. Such as animals, plants, some fungi, but also some single-celled organisms as well including algae that live in the oceans. These are the 3 domains of life, and they break up into smaller subdivisions of organisms. We call the bacteria and archaea these single-celled organisms, the prokaryotes. Prokaryotes is a general name that's quite useful to classify micro-organisms, bacteria, and archaea. It's a term one frequently sees when describing simple single-celled organisms. So how do we build this phylogenetic tree, this tree of life? Well, one way in which we can do this is just to compare the shapes of different organisms.

Most of us can tell the difference between a horse and a dog for instance. So that might be one way in which we could classify organisms, this is called taxominay. Another way in which we can do this is to look at molecules. Inside organisms and compare them. And the reason why this is useful is that it's sometimes very difficult to tell the difference between two organisms just based on their shape. This is particularly true with microorganisms. Here are a couple of images of microbes. On the left, you can see Escherichia coli. A typical microbe grown in many microbiology labs and used to study is the biochemistry of microbes for instance. On the right, you can the vibrio cholera, the microbe responsible for cholera. If you look at these two microbes under a microscope, they look very similar. They're both rod-shaped. They're both about the same sort of size. It would be very difficult to build a tree of life by looking at all the microbes in the world and comparing their shape. So, we have to use more sophisticated methods to tell the difference between different organisms. And one way in which we can do this is to look at molecules that are very important for fundamental cell processes. We saw earlier in this lecture course how cells need to replicate their DNA, the information storage system, and also read that DNA reads the instructions in the DNA, to carry out basic cell functions. There's a particular molecule called ribosomal RNA that's responsible for translating DNA into proteins, for reading the DNA code, and for telling you that it's a protein. And in fact, reads the code on RNA, and ribonucleic acid. But because this ribosomal RNA is so crucial to cell function, it hasn't changed much over billions of years. So, if we look at two organisms that are very closely related, we find that ribosome RNA has not changed much since those two organisms diverged, in the evolutionary record. If we look at ribosomal RNA from two organisms that diverged a long time ago, that split apart in the evolutionary record, a much greater time ago, then we find that the ribosomal RNA has changed to a much greater extent. And so, by looking at ribosomal RNAs in different organisms and how much it's changed, we can start to build up a tree of the evolutionary distance between those organisms and so these molecular methods allow us to build a tree of life. Another reason for looking at molecular structures is that many microorganisms cannot be cultured. In the laboratory, this is called the great plate count anomaly. And the reason for this is we really don't know how to grow these microbes from the natural environment. So, if we really want to build a true tree of life, a tree of life that represents all of life on Earth, going about trying to grow all these things in a laboratory can be very laborious and not very productive. It's much easier to extract these molecules from the environment and identify them without even growing those microbes in the laboratory. And if we do that, we find that some places on the Earth, such as deserts actually, turn out to be quite diverse in terms of the microbiology, and the microbes that they harbor. So, using these molecular methods, like ribosomal RNA, not only allows us to build a tree of life and to tell the evolutionary distances between organisms, but it also allows us to get a better grasp of the full diversity of life on our planet and what that common ancestor might have been like. Well, so far, we've discussed the tree of life that assumes that genetic material is just transferred from parent to offspring one generation after another. It turns out that it's not quite that simple. Microbes can do remarkable things. If I told you, for instance, that you could walk up to a person in the street and just by touching them your eye color would become the same as their eye color, you would think that's a ridiculous idea and some sort of weird science fiction idea. But in fact, bacteria can do exactly that. In the environment, they can take up DNA from other bacteria and they can change their characteristics by absorbing this DNA. And one process by which they do this is called conjugation whereby they directly transfer DNA from one bacterium to another. This process makes the tree of life a little bit more complicated.

Because genetic information is not just transferred vertically, from one generation to another through time, but the material can also be transferred horizontally as we say, between species, at any particular point in time. And this means the tree of life is less like a tree of life and more like a web of life, or a ring of life. Or perhaps more amusingly, some people have referred to it as a shrub of life. But it's certainly more complicated than just branches. In branching off at different points in the history of life on Earth, we have to understand this horizontal gene transfer between organisms, particularly microorganisms in order to understand how that relationship between organisms has developed over time. So what have we learned? We've learned that phylogenetic studies the evolutionary relationship between groups of organisms. The Tree of Life can be divided into three domains. They are bacteria, archaea, and eukaryotes, including humans. Prokaryotes which broadly include bacteria and archaea are the oldest and most abundant of all organisms. Most prokaryotes are known only through their genetic material, the genetic sequences. The tree of life remains debated and alternative phylogenetic models have been proposed. But this is yet another task of biologists to understand what the earliest common ancestors to life on Earth were. So the question asked is, how the phylogenetic tree developed and the relationship between the organisms on the Earth. (Cockell 2013)

CHAPTER 4
LIFE IN EXTREME CONDITIONS
4.1 Early Earth Conditions

How has life managed to survive over 3 billion years on the Earth? And what adaptations does it need to survive environmental changes throughout its long tenure on the Earth? One of the best ways to answer this question is to go and look at how life manages to survive and grow in extreme environments. What do we mean by extreme? Some people think that it's a very anthropocentric term and is just a matter of taste. What may be extreme to one organism is not extreme to another. But we think that there is something much more fundamental to it than that.

Biological systems do only function in a continuum of particular physical and chemical extremes. And as we go to these so-called extremes, such as high temperature, low temperature, and increased pressure, we find that the diversity of life generally in these environments tends to decrease. So, it does seem that there are real extremes, extremes beyond which life cannot go. Boundaries to the biosphere are determined by physical and chemical extremities. And studying the organisms at these extremities can tell us much about how life on earth has managed to cope with conditions on our planet, and just as interestingly whether life might be able to cope with the physical and chemical extremes to be found on other planetary bodies. The organisms that inhabit these extreme environments we call extremophiles are literally extreme lovers. They're mainly the prokaryotes, the bacteria, and archaea we saw in the tree of life but here are some eukaryotes that live in extreme environments. Even flies, for example, live at the edges of volcanic hot springs in Yellowstone National Park. But by and large, once we get to really great extremes on the earth, we find that the organisms that are inhabiting these environments are the prokaryotes. What are the types of extreme environments that exist out there? Well, these are some of the environments that we'll look at in a little more detail. There are hot and cold places. There are places that are very salty, or very dry. There are also places that are very acidic or very alkaline or basic.

And there are also places with intense pressure and also intense radiation. Let's have a look, to begin with, at a hot environment. There are many hot environments on the earth that harbor life. A good example is deep-sea hydrothermal vents. Places where reduced fluids containing sulfides and other types of chemicals are spewing out from the crust of the earth into the oceans. And these hydrothermal vents can be producing water well over a hundred degrees centigrade because of the high pressure in the deep oceans. The organisms that inhabit these hot environments also include volcanic hot springs in places like Yellowstone National Park are called thermophiles. If they grow between 50 and 80 degrees or if they grow at really high temperatures above 80 degrees, they're called hyperthermophiles. Now it's important to remember that these organisms aren't just capable of growing at these temperatures, they actually need to grow at high temperatures. If you take a hypothermophile, for example, whose optimum growth temperature is above 80 degrees, and you bring it down to room temperature, it will generally die. These are microbes that actually have to be growing at these very high temperatures. A good example is Methanopyrus kandleri. This is an Archaea that inhabits black smokers, and deep ocean hydrothermal vents that are black because they're producing sulfide minerals that are produced in the oceans as these fluids gush out of these black smoker vents.

The organism can grow up to temperatures of 110 degrees Centigrade. Some of the challenges it faces, like all thermophiles or hyperthermophiles are the breakdown of biomolecules. These high temperatures impart energies to the microorganisms that tend to cause the biomolecules to break down. They also have a problem with membrane fluidity. At very high temperatures, the energy causes the membranes to start shaking apart, almost quite literally. And that fluidity in the membrane can cause problems for the integrity of the cell membrane. How does it deal with the challenges of living in these high-temperature environments? Well, two of the ways it deals with this, is by evolving thermostable proteins and enzymes. Enzymes are biological catalysts involved in carrying out chemical reactions in the cell. These proteins have extra chemical bonds and other types of features that maintain their stability at these very high temperatures. It turns out that these proteins or catalysts, these enzymes have, have commercial uses as well. For example, thermostable catalyst enzymes are used in biological washing powder. One of the reasons why your washing powder can work at high temperatures is because it contains proteins or enzymes from microorganisms that have been isolated from hydrothermal springs and volcanic hot springs. So you see we can use these thermal stable enzymes and proteins for some very prosaic but commercially useful applications. Other adaptations include changes to the cell membranes and composition to allow those membranes to maintain stability at high temperatures.

Microorganisms have also been found in freezing environments such as the depths of Antarctic ice sheets. This is an example of a deep lake in the Antarctic ice sheet called Vostok, and just above that lake are microorganisms in the ice, in the accretion ice that forms, above that deeply buried lake. In these sorts of ice sheets, microorganisms adapted to cold conditions can grow, and these are called psychrophiles, microorganisms that can grow at temperatures of less than 15 degrees Centigrade. What are the challenges of living in a cold environment? Well, one challenge of course is membrane damage from ice. If ice crystals form in the cell, it can damage the membranes. Another problem is decreased membrane fluidity. In the very cold temperatures, the lipids literally begin to solidify a bit if you put butter in your fridge it begins to get a lot more solid. Those fatty acids, those lipids in the cell membranes begin to solidify at very cold temperatures and reduce the fluidity of the membrane that's necessary for the cell to export materials and import nutrients. There is an obvious problem in freezing environments if the availability of liquid water. Much of it is frozen up in ice. And so, organisms can have trouble getting hold of liquid water that they need to carry out.

4.2 Biochemical Reactions

So how is it that these microbes can adapt to these extremely cold conditions? Well, one way in which they adapt is by altering their membrane composition, a bit like the microbes living at high temperatures. They need to change the composition of the membranes to maintain their fluidity at low temperatures.

One way in which they can do this is to incorporate more unsaturated fatty acids into their membranes. These unsaturated fatty acids have kinks in the membrane structure that pulls, pushes apart the membrane and makes it more fluid under cold conditions. Some of these microbes and other organisms also have anti-freeze agents such as sugars. And these sugars prevent ice crystals from forming in the cells. And reduce the chances of ice crystals falling and damaging the membranes. Another way in which they can circumvent the problems of growing at very low temperatures is in both hot and cold environments. What about salty and dry environments. Well, microorganisms have been found that can inhabit very salty environments such as microbes that live in the Dead Sea, and in deposits of salt around the Dead Sea. These are called halophiles, and these halophiles, literally salt lovers, can grow in salt concentrations between 15 and 37%. Microbes live in deserts, also capable of tolerating very dry conditions, and these are called xerophilic microbes. You can find them in the deserts of the Atacama, the Sahara, and other dry and desiccated environments of the Earth. The challenges of living in salty and dry conditions are actually quite similar. One problem is osmotic pressure and water availability in very high salt concentrations, the salt. Has the tendency to pull water out of the cells by the process of osmosis, so cells have a problem in hanging on to their liquid water. And of course, in dry deserts the problem is also the water tends to evaporate, which causes problems for the maintenance of water in the cell, also osmotic effects in very dry conditions. Another challenge faced by microbes in both salty and dry environments is membrane integrity, maintaining the integrity of the membranes, and preventing them from falling apart, under these very high salt concentrations that tend to disrupt biomolecules. And under very dry conditions, where the dry conditions also tend to cause disruption to the membranes, lack of water. How do organisms adapt to these very extreme, salty, and dry conditions? Well, two ways they can adapt are to control water loss from the cells.

They can produce salts and other solutes within the cell that tend to hang on to the water make it more difficult for it to dissipate from the cell. Another way in which they can deal with these extreme conditions is to go into a state of dormancy. When they're dormant, they're not active, but it allows them to wait around until conditions improve is a particular case for microbes living in deserts, where they may want to go dormant and wait until liquid water becomes available and they can reproduce and grow again, and then go into a stage of dormancy when it dries up. Microbes have also been found living in extremes of pH, for example the Rio Tinto River in Spain has pHs down to 0.4. High concentration of protons in the water makes it very, very acidic. There are also places with very high pH such as Mono Lake in the United States that has a pH up to 12.5. These are very alkaline or basic environments that pose great challenges to microbes. The challenges that are faced to the organisms that live in these environments include the breakdown of their cellular components for example in acidic environments. Very high proton concentration that's responsible for that acidity can disrupt biomolecules, causing them to become inoperative. This very challenging pH conditions are also a great change to metabolic processes, processes for gaining energy and carrying out chemical reactions inside the cell. How do cells adapt to these extreme pH's? Well one way in which they can adapt is to. Regulate the pH inside the cell in order to make it neutral. So, for example, a microorganism living in very acidic environments will pump out protons from the cell to maintain the inside of the cell at near neutral pH. And at neutral pH the biomolecules are much more stable and metabolic processes can proceed without disruption. In other words, these cells really don't like to be in acidic environment, but they can change their internal conditions such that they can survive and grow in acidic environments without the acid on the outside of the cell effecting cellular reactions inside the cell. This is a really ingenious way by which cells can survive and grow in acidic conditions. Similar sorts of processes are also found in very alkaline environments.

Organisms have also adapted to life under high pressure. In fact, many of the habitats on the Earth for life are actually at high pressure. Two examples are the deep oceans and the deep crust of the Earth. In the deep oceans, many of the trenches are at great depths, such as the Mariana Trench at 11 kilometers depth. And here pressures exceed 1000 atmospheres. The microbes that can live in these environments are called Piezophilic microbes, literally, pressure-loving microbes (They adapt well to…conditions). In fact, at the current time, we don't know what the upper-pressure limit for life is. Certainly, they can survive at high pressures to be found in these deep ocean trenches. Microbes are also found in the deep crust living in rocks kilometers underneath the surface of the earth. The challenges of living at high pressure the pressures cause the tight packing of molecules and a loss of fluidity in cell membranes. Pressure can also cause impaired cellular functions and activities, particularly of enzymes that are necessary to carry out the catalytic functions of chemical reactions inside the cell. How do cells and organisms adapt to live in these high-pressure environments in the deep oceans or in the crust? Well, one way in which they can adapt is by changing gene expression. They can produce, for example, molecules that enhance the uptake of nutrients and other elements that they need for growth across the cell membrane and thereby adapt to these challenging environments under high pressure. They can also change their membrane structure, for example by introducing unsaturated fatty acids to increase membrane fluidity under high pressures. And you'll notice that unsaturated fatty acids were also the way in which microbes can adapt to low temperatures. So, some of the mechanisms that organisms use to adapt to one extreme are also used to adapt to other extremes. They can be common responses to different environmental extremes. We might think about one of the most extreme environments known to man and that is the conditions of outer space, can micro-organisms survive in outer space. Space is characterized by extremes of radiation, freezing temperatures, desiccation, and no oxygen. A few years ago, my own laboratory launched rocks into earth's orbit and these rocks were bolted onto the outside of the International Space Station. And we brought them back to earth a year and a half later to see whether anything had survived in space.

And we found a single microorganism, a Gloeocapsa which is a type of cyanobacterium, a photosynthetic micro-organism that was capable of surviving in the extreme conditions of space, for a full year and a half. Of course, it didn't grow in space, but it did survive. These types of experiments show how outer space is an environment that can even be survived by some microorganisms showing how hardy they are and how able they are to resist environmental extremes, if only for a short length of time. Many extremes that we find on the Earth don't occur in isolation. I've talked about high and low temperatures, I've talked about salty environments and highly acidic and alkaline environments, but in fact, in many natural environments, there are frequently multiple extremes. And biologists are very interested in polyextremophiles, extremophiles that can tolerate multiple extremes. Here's just one example, a rather famous organism called Deinococcus radiodurans. This is a microbe that can tolerate high levels of radiation. It's also found in environments with cold temperatures. It's found in deserts that are very dry. Some of these organisms can tolerate vacuum conditions and members of this group, the meningococci, are also found in very acidic environments as well. So, we see how there are polyextremophiles that can survive multiple extremes and by studying these microorganisms we can get a better understanding of how microbes tolerate the boundaries of extremes to find the boundaries of the Earth's biosphere. Why is this of any interest to biologists? Well, there are really two reasons why we're interested in studying microbes at extremes and particularly those microbes. It can tolerate multiple extremes. First of all, of course, it tells us about the boundaries of the earth's biosphere. What are the limits of life on earth? When do we go beyond those limits? And how might environmental changes, throughout the history of the earth, even environmental changes caused by humans affect life on Earth and those microorganisms then inhabit the outer boundaries of the biosphere. But the other reason for being interested in studying life in extreme environments is because it might give us better ideas about the habits and abilities of other worlds, such as Mars, Europa, Enceladus, and Titan, and other planetary bodies of interest to astrobiologists. Once we know the physical conditions of those planetary bodies, we want to be able to assess whether they are within the boundaries of life and whether life might be able to persist on those planetary bodies.

The only way in which we can do that is by studying microorganisms and life in general in extreme environments and seeing whether the extremes that we observe on other planetary bodies are extremes that can be tolerated by lifeforms that we know on the Earth. So, the study of life in extreme environments is essential for assessing habitability, and the ability of other planetary bodies to harbor life. So, what have we learned in this lecture? We've learned that extremophiles are organisms that tolerate or require physical and chemical extremes in order to be able to survive or grow or reproduce. We've learned that although extremophiles span all major domains of life, most of them come from the prokaryotes, bacteria, and archaea. We've learned that adaptations to extreme conditions sustain cell integrity and function frequently via modifications to the cell membrane and molecules, such as proteins and enzymes, and catalytic molecules in cells. We've learned that interactions between multiple extremes may influence extremophile growth and survival. The study of life in extreme environments is pivotal to understanding the emergence of life on the Earth and understanding the boundaries for life in the universe at large. And finally, we've also learned that molecules isolated from extremophiles may also have commercial and industrial use.

4.3 The Rise of Multicellularity
- Life through Time

There have been many changes in life over its long tenure on the Earth. But undoubtedly one of the most remarkable changes was the development, of multicellularity. What is a multi-cellular organism? Well, a multi-cellular organism is generally one made up of cells that have irreversibly differentiated or altered to carry out specialized functions. For example, in us we might have cells that are growing into liver cells, some cells will become skin cells, and cells that do other functions. This different diversity of specialized cells comes together to make the whole human organism.

This is generally what multi-cellular organisms are. At some point in the history of life on Earth, unicellular organisms, or microorganisms transition to becoming multi-cellular. How did this happen, and when did it happen? For the first 2.5 billion to 3 billion years of life on Earth, there were only micro-organisms. This evidence is preserved in the early rock record in controversial micro-fossils and chemical signatures. These remnants of early micro-organisms become better preserved as the rock record improves through time. And then between 585 million years ago and 542 million years ago, fossils begin to appear in the rock record that suggests early multicellular organisms. What is the earliest evidence for these multi-cellular organisms? Well, some of the 1st records of multi-cellular creatures. Ediacaran Fauna: These are fossils preserved in rocks between 585 and 542 million years ago. They're very enigmatic creatures, strange, tubular, front-shaped organisms which lived during this period. And many people think that these were experiments in early body plans. Experiments that would eventually give rise to the body plans of multi-cellular organisms That we're familiar with today. These were first discovered in Australia. And this is an artist's impression of these creatures and what they may have looked like many millions of years ago. How did multi-cellularity arise? Well, at some point, single-cell microorganisms. Made the transition to multi-cellular organisms. We don't quite yet know how that happened. It's one of the great questions in biology, to fathom, how multi-cellularity evolved. There are two possible explanations. One is an internal change in cells, and one is possible environmental change.

The internal change may have occurred by some alteration or mutation in the genetics. At some point when cells were dividing, they divided in such a way as to differentiate some change in the genetic Code caused cells to change their functions but remain in communication with one another such that a mass of previously almost identical single-celled micro-organisms became a mass of differentiated cells, an early multi-cellular organism. But one would still have to question why that happened. What triggered this genetic differentiation within cells, this emergence of multi-cellular features? Another explanation is the rise of oxygen in the earth's atmosphere. Oxygen makes possible aerobic respiration whereby organic material is essential to burn in oxygen to produce energy. Aerobic respiration is a very good way of generating large quantities of energy and it allows the. Things like running and jumping, and even running a brain. So aerobic respiration may have been key to multi-cellular organisms. The rise of oxygen in the earth's atmosphere to similar levels to today may have allowed large,
complex organisms to evolve and develop. Once it did evolve, why did it remain around? Why was it selected?

As Charles Darwin said, it is not the strongest of the species that survives, nor the most intelligent that survives. It is the one that is most adaptable to change. Multi-cellularity may have been advantageous in several ways. For example, the increased size would have allowed organisms to escape predation and competition from other multi-cellular organisms. So, once it did emerge, it persisted because it was necessary for organisms to escape already existing, multi-cellular organisms. Multi-cellularity may also have allowed organisms to exploit new ecological niches. Regions where there were resources or food supplies perhaps not being efficiently used by other types of life, allowing life to expand into new physical and ecological territories. Cellular specialization is also quite energetically efficient. The cells specialize in one particular function in an organism. They can do that function very efficiently and very well and thereby direct their energy to that particular function. Multi-cellularity may also have allowed for physical protection from the external environment and protection from other predators, but also protection from physical extremes. There were also computational advantages. Multi-cellular organisms were able to develop much more complex behavioral characteristics, which would have allowed them to have competed more effectively.

To have hunted out new territories and new food supplies and new resources more efficiently, we have to be careful in interpreting these changes as necessary for organisms to compete. We know that many microorganisms are energetically very efficient. We also know that they can survive in extreme environments very well. And also compete with other microorganisms. So the reasons that multicellularity both evolved and persisted in the environment are still controversial. This is one of the questions to be addressed by biology. One of the more remarkable transitions in the fossil record occurs. 542 million years ago, when we start to see fossils appearing in the rock record that represent a huge diversity of multi-cellular organisms. This period in the rock record is called the Cambrian Explosion. One of the reasons why the diversity of multi-cellular organisms may greatly increase in rock record of this time is thought to be because organisms began to evolve the ability to precipitate minerals. Used for skeletons and shells. And once these skeletons and shells became preserved, we began to see a greater diversity of life. In the e rock record, before those multicellular organisms were in general soft body and those soft bodied organisms are not well preserved in the rock record.

So, the Cambrian explosion, this increase in the diversity of multi-cellular organisms in the rock record, maybe an artifact of preservation, but nevertheless it shows us that this is a period of Earth history where the diversity of multi-cellular organisms including their ability to be preserved, was greatly increased. I have given you, perhaps, some idea that life on Earth is rather continuous. We've gone from unicellular microorganisms to multi-cellular life, right the way through to the present day. But it's important to remember that multi-cellular organisms, as probably with micro-organisms, have been subject to mass extinctions in their part. Since the rise of multi-cellular life, we recognize five big mass extinctions in the rock record. They have different causes. And the causes of these mass extinctions are very controversial, another area which is a very active field of investigation in both biology and astrobiology. For example, the cretaceous Paleogene extinction 65 million years ago is thought to have been caused by an asteroid or comet impact. Other mass extinctions have probably been caused by changes in the environment other than asteroid and comet impacts, for example from volcanic eruptions.

All of these causes are a matter of debate. But after each extinction, biodiversity on the Earth has always recovered or increased, showing how resilient life on earth is to these large environmental changes that are capable of causing extinctions. Extinctions may not always be bad for all life, because of course extinctions create spaces for new organisms that survive to flourish and emerge into empty habitats. For example, the destruction of the dinosaurs 65 million years ago allowed mammals to dominate the surface of the earth, eventually leading to human beings. So, extinctions are a double-edged sword. They result in the destruction of large swaths of the diversity of life on earth, but they also open up new opportunities for life, to expand into those empty niches and habitats. So, what have we learned in this lecture? Hopefully what we've learned, is that unicellular organisms prevailed for about the first 3 billion years of life's history. Fossils from about 585 to 542 million years ago seemed to comprise the first evidence for multi-cellular life at least in experiments in early body plants. Multi-cellularity arose maybe as a result of interactions between the cells, genetic changes within organisms themselves, or changes in atmospheric conditions, such as a rise of atmospheric oxygen. The advantages of multi-cellularity may have included an increase in size and specialization and physical protection from the environment. And even the conditions for the development of more complex behavior, which would have enhanced their ability to compete in the environment. The Cambrian Explosion, as recorded in the rock record, shows a great diversification of life, including the appearance of hard-bodied organisms. And finally, mass extinctions on Earth have been followed by periods of biological recovery and diversification. But extinctions have very much been a fact of life for multi-cellular life since it first arose.

4.4 The Great Oxidation Event
~ Life through Time

We look to conditions on the early Earth, but the conditions on our planet have not been constant since the emergence of life through to the present day. In fact, many environmental changes have occurred during that long period of time. We could look at a number of these but let's just focus on one of them that is of great interest to biologists. And it's called the great oxidation event. It's that period in earth's story where oxygen raised to levels that we're familiar with today. In fact, it occurred over more than one period.

The rise in oxygen turns out to be very important for life because oxygen enables aerobic respiration, a very efficient way of gathering energy. Where did the oxygen come from? Well, the early earth had less than 0.1% of the oxygen we have present in our atmosphere today. And we'll look at the evidence for that in a little while. It raises several questions of great interest in biology and here are some of them. When did oxygen become abundant? And what was the cause of the rise of oxygen? How do we actually track the rise of oxygen? How do we know it increased and how do we find out when it increased? And how did the rise in oxygen affect conditions on the Earth, particularly conditions for life? Let's look at some of the ideas, and answers to these questions. First of all, when did what, oxygen rise, and what were the periods of the increase in oxygen in the Earth's atmosphere? This is a graph showing the concentration of oxygen over time. Now on the x-axis at the bottom there, we go from the present day on the right back to the formation of the Earth over 4 billion years ago on the left. And you can see the red line tracks oxygen through time. In the early history of the Earth, oxygen was less than 0.1% of the present. And we can tell that by what we call proxies. These are geochemical signatures in the rock record, types of rocks that can tell us that the oxygen was within certain bounds, we'll come back to that evidence in a bit. And then around 2.4 billion years ago, oxygen rises to something like 5 to 18% of present levels. And it stabilizes there for a long period of time. And then between about 600 to 800 million years ago, there was a second rise in oxygen that leads to the oxygenation of the oceans and leads to oxygen concentrations we're familiar with today, around 21 percent, although it has varied during that time.

4.5 What caused the great oxidation events?

Particularly, the first one was 2.4 billion years about when oxygen rose from very low levels to something of the order of 5 to 18% of the present day. Well, first of all, we have to understand both the sources of oxygen and also the sinks for oxygen, the places where oxygen disappears, and how it's removed from the atmosphere. The source of oxygen is photosynthesis. For example, plants and cyanobacteria are microorganisms that produce oxygen as a result of photosynthesis. Oxygen is a waste product of photosynthesis, the metabolic process by which organisms get energy from the sun to build complex carbon molecules. Photosynthesis, at least oxygenic photosynthesis, produces oxygen as a waste product. So that's the main source of oxygen. That's how oxygen would have been pumped into the atmosphere- from photosynthesis. How does oxygen get removed from the atmosphere? There are really two main mechanisms by which this happens. The first is that when organisms die their organic compounds literally react with oxygen, consuming oxygen in the atmosphere. The first organisms are buried before they can react with the oxygen. Then they might have a chance to remove oxygen from the atmosphere, and there's a net buildup of oxygen in the earth's atmosphere. So, the burial of organic material, the burial of dead organisms, can help oxygen build up in the atmosphere. Another way in which it can be removed from the atmosphere is by reacting with volcanic gases. For example, oxygen can react with Hydrogen Sulfide or Methane and in the process get removed from the atmosphere and resulting in lower oxygen concentrations. So, if we want to build up oxygen in the atmosphere, one of two things has to happen. Either less of it must react or be taken out of the atmosphere, or more of it has to be produced. So why did we get this abrupt rise in oxygen about 2.45 billion years ago? Well, some possible explanations are that there was a sudden decrease in the elements or gases that react with oxygen being produced, for example, by volcanic eruptions. This would have led to a net buildup of oxygen in the atmosphere. Another explanation is that photosynthesis evolved about two point four five billion years ago and produced oxygen that builds up in the atmosphere.

However, the record and the following genetic evidence show that photosynthesis is much more ancient than two point four five billion years. So, we think that photosynthesis was occurring long before the great oxidation event, but that doesn't look like a particularly good reason. Another reason could be an increase in the burial of dead organisms. Less of the organisms were around to react with oxygen in the atmosphere and remove it. And as a result, oxygen built up in the atmosphere. It could be that some of these conditions occurred in combination. And that there was a switch between two different stables conditions.

A stable condition in the Earth environment where there is very low oxygen, and another stable condition where there's high oxygen. And at some point, 2.45 billion years ago, the Earth flipped between those two stables conditions. However, it occurred, whatever the reasons for it, it had profound consequences for the earth's environment. Arising oxygen would have mucked up methane and methane is a greenhouse gas. So as oxygen rose and it mucked up this greenhouse gas, temperatures on the Earth would have reduced and there would have been a period of cooling. In fact, there is evidence for glaciations following these great oxidation events. The production of oxygen also would have changed the chemical nature of elements on the Earth's crust and so changed the chemical cycling of carbon, nitrogen, sulfur, iron, and other elements, essential as nutrients for life. It would have radically changed the biosphere and the accessibility of nutrients to supply the biosphere. For life, the increase in oxygen was a double-edged sword. Of course, for those microbes that were living on the earth before the rising oxygen, they were adapted to anaerobic conditions, conditions without oxygen. The rising oxygen would have been fatal for many of these organisms. And they would have had to have persisted or been confined to anaerobic environments, such as, for example, deep ocean hydrothermal vents at the bottom of the oceans or other areas for example in the crust, where reducing conditions would have maintained anaerobic or oxygen-free conditions. But the other side of this is that oxygen would have allowed for aerobic respiration.

Aerobic respiration is about 16 times more efficient as a method of generating energy than anaerobic means of growth. It allows for complex behaviors, even running a brain, and so the production of oxygen in high concentrations in the atmosphere may have led to multi-cellularity. It may have led to organisms that we're able to use oxygen as a way of gathering large quantities of energy.

In order to grow and reproduce another way in which we can look at past oxygen levels and try to understand when it arose, was to look at finer genetics, to try and understand when the first organisms appeared that might have been responsible for this rise in oxygen. For example, cyanobacteria, once called blue-green algae, are a group of organisms that are photosynthetic and produce oxygen. We now know from phylogenetics that cyanobacteria were very deep branching, and that means they appeared on the Earth a long time ago, probably a lot longer before the great oxidation event. And they would've been responsible for producing oxygen in the earth's atmosphere. Indeed, they might explain why there appeared to be episodes of oxygen production even before the great oxidation event. Photosynthetic micro-organisms might have been producing oxygen well before three billion years ago. But this oxygen they were producing was being mopped up, for example, by gases, reduced by volcanic eruptions long before the great oxidation event. Another way in which we can track oxygen in the rock record is to look for particular minerals that are only produced at low oxygen concentrations. Here are just some examples. For example, pyrite is iron sulfite, it's a mineral that's only produced at oxygen concentrations much less than <0.1%Present Atmospheric Level (PAL). Uraninite is a mineral of uranium, uranium oxide, that's only produced at oxygen levels much less than 0.01%. Present Atmospheric Level. And Siderite, which is iron carbonate, is a mineral that generally is only produced at oxygen concentrations much less than 0.001% atmospheric levels. So, if we find these minerals in the rock record, it tells us that oxygen levels were very low at that period of Earth's history. This is how geologists have managed to track the history of oxygen throughout Earth's history, by looking at these minerals in the rock record and by inferring the levels of oxygen that must have existed - very low levels for these minerals to have been able to form.

Another way in which we can trace oxygen throughout history is by looking at elements that change their solubility in water. The amount they're dissolved in water depends upon the amount of oxygen in the environment. And the depletion of certain elements from ancient soils or paleosols can be used to infer the oxidation state of the atmosphere. For example, under very low oxygen conditions Iron is very soluble in water. And generally, in rocks before or about 2.4 billion years ago, we see very low levels of iron, and an increased level of iron after about 1.8 billion years ago, when the earth was more oxygenated. Some of the most remarkable features in the earliest rock record are banded iron formations. These are thick geological deposits containing a higher content of iron oxide. It's essentially rust. How did they fall? But one theory is that the early oceans contained very high concentrations of dissolved iron, which would be the case if the ocean had very little oxygen in it. In an environment that was essentially oxygen-free. Every now and again, this iron would have been circulated through the oceans, and it may have come into contact with more oxidized surface waters in the ocean. The iron would have oxidized. It would have settled out in the oceans and formed these thick geological deposits. These banded iron formations are widespread only in the rock record before about 2.4 billion years ago, before the great oxidation event. But they essentially disappear after this time. And so, these banded iron formations are evidence that high concentrations of dissolved iron were caused by low oxygen conditions on the early earth. So, what have we learned in this lecture? What we've learned is the levels of oxygen rose rapidly about 2.4 billion years ago in the great oxidation event. The cause of this rapid shift is quite ambiguous but nevertheless, however it occurred, the rising oxygen had a significant effect, probably on global climate as it eliminated the methane greenhouse effect. The increased oxygen is thought to have facilitated the rise of complex multi-cellular organisms. And finally, the interpretation of the rock record is really key to our understanding of the history of oxygen in the Earth's atmosphere. (Cockell)

CHAPTER 5

EVOLUTION BY NATURAL SELECTION

5.1 Natural Selection

In 1838, Charles Darwin discovered the principle of evolution by natural selection and revolutionized our understanding of the living world. Darwin was 28 years old, and it was just two years since he had returned from a five-year voyage around the world as a naturalist on the HMS Beagle. Darwin's observations and experiences during the journey had convinced him that biological species change through time and that new species arise by the transformation of existing ones, and he was avidly searching for an explanation of how these processes worked. In late September of the same year, Darwin read Thomas Malthus' Essay on the Principle of Population, in which Malthus argued that human populations invariably grow until they are limited by starvation, poverty, and death. Darwin realized that Malthus' logic also applied to the natural world, and this intuition inspired the conception of his theory of evolution by natural selection. In the intervening century and a half, Darwin's theory has been augmented by discoveries in genetics and amplified by studies of the evolution of many types of organisms. It is now the foundation of our understanding of life on Earth.

The profound implications of evolution for our understanding of humankind were apparent to Darwin from the beginning. We know this today because he kept notebooks in which he recorded his private thoughts about various topics. The quotation that begins this prologue is from the M Notebook, begun in July 1838, in which he jotted down his ideas about humans, psychology, and the philosophy of science.

In the 19th century metaphysics involved the study if the human mind. Thus, Darwin was saying that since he believed humans evolved from a creature something like a baboon, it followed that an understanding of the mind of a baboon would contribute more to an understanding of the human than would all of the works of the great English philosopher John Locke. Darwin's reasoning was simple. Every species on this planet has arisen through the same evolutionary processes. These processes determine why organisms are the way they are by shaping their morphology, physiology, and behavior. The traits that characterize the human species are the result of the same evolutionary processes that created all other species. If we understand the processes, and the conditions under which the human species evolved, then we will have the basis for a scientific understanding of human nature. Trying to comprehend the human mind without an understanding of human evolution is, as Darwin wrote in another notebook that October, —like puzzling at astronomy without mechanics.‖ By this Darwin meant that his theory of evolution could play the same role in biology and psychology that lead Isaac Newton's laws of motion had played in astronomy. For thousands of years, stargazers, priests, philosophers, and mathematicians had struggled to understand the motions of the planets without success. Then, in the late 1600s, Newton discovered the laws of mechanics and showed how all of the intricacies in the dance of the planets could be explained by the action of a few simple processes.

Figure 5-1: Darwin at age 29

In the same way, understanding the processes of evolution enables us to account for the stunning sophistication of organic design and the diversity of life, and to understand why people are the way they are. As a consequence, understanding how natural selection and other evolutionary processes shaped the human species is relevant to all of the academic disciplines that are concerned with human beings. This vast intellectual domain includes medicine, psychology, the social sciences, and even the humanities. [xx]

5.2 Adaptation by selection - Darwin's Theory

Even the casual observer can see that organisms are well suited to their circumstances. For example, fish are clearly designed for life under the water, and certain flowers are designed to be pollinated by particular species of insects. More careful study reveals that organisms are more than just suited to their environments – they are complex machines, made up of many exquisitely constructed components, or adaptations, that interact to help the organism survive and reproduce. Charles Darwin was born into a well-to-do, intellectual, and politically liberal family in England. Like many prosperous men of his time, Darwin's father wanted his son to become a doctor. But after failing at the prestigious medical school at the University of Edinburgh, Charles went on to Cambridge University and resigned to become a country parson (clergyman). He was, for the most part, an undistinguished student, and was much more interested in tramping through the fields around Cambridge in search of beetles than in studying Greek and mathematics. After graduation, one of Darwin's biology professors, William Henslow, provided him with a chance to pursue his passion for natural history as a naturalist on the HMS Beagle.

The Beagle was a Royal Navy vessel whose charter was to spend two to three years mapping the coast of South America and then return to London, perhaps by circling the globe. Darwin's father forbade him to go but preferred that he get serious about his career in the clergy, but Darwin's uncle (and future father-in-law) Josiah Wedgwood intervened. The voyage turned out to be the turning point in Darwin's life. His work during the voyage established his reputation as a skilled naturalist. His observations of living and fossil animals ultimately convinced him that plants and animals sometimes change slowly through time and that such evolutionary change was the key to understanding how new species came into existence. This view was rejected by most scientists of the time and was considered heretical by the general public.

Figure 5-2 : Darwin at age 51

5.3 Darwin's Postulates

Darwin's theory of adaptation follows from three postulates: the struggle for existence, variation in fitness, and the inheritance of variation.

1. The ability of a population to expand is infinite, but the ability of any environment to support populations is always finite.
2. Organisms within populations vary, and this variation affects the ability of individuals to survive and reproduce.
3. The variations are transmitted from parents to offspring

Darwin's first postulate means that populations grow until they are checked by the dwindling supply of resources in the environment. Darwin referred to the resulting competition for resources as —the struggle for existence.‖ For example, animals require food to grow and reproduce. When food is plentiful, animal populations grow until their numbers exceed the local food supply. Since resources are always finite. It follows that not all individuals in a population will be able to survive and reproduce.

According to the second postulate, some individuals will possess traits that enable them to survive and reproduce more successfully (producing more offspring) than others in the same environment. The third postulate holds that if the advantageous traits are inherited by offspring, then these traits will become common in succeeding generations. Thus, traits that confer advantages in survival and reproduction are retained in the population, and traits that are disadvantageous disappear.

When Darwin coined the term natural selection for this process, he was making a deliberate analogy to the artificial selection practiced by animal and plant breeders of his day. A much more apt term would be —evolution by variation and selective retention.

5.4 An example of adaptation by Natural Selection

In his autobiography, first published in 1887, Darwin claimed that the curious pattern of adaptations he observed among the several species of finches that live on the Galapagos Islands off the coast of Ecuador – now referred to as Darwin's finches – was crucial in the development of his ideas about evolution. Recently discovered documents suggest that Darwin was actually quite confused about the Galapagos finches during his visit, and they played little role in his discovery of natural selection. Nonetheless, Darwin's finches hold a special place in the minds of most biologists.

5.5 The Evolution of Complex Adaptations

The example of the evolution of beak depth in the medium ground finch illustrates how natural selection can cause adaptive change to occur rapidly in a population. Deeper beaks enabled the birds to survive better, and deeper beaks soon came to predominate in the population. Beak depth is a simple character, lacking the intricate complexity if an eye. However, as we will see, the accumulation of small variations by natural selection can also give to complex adaptations.

Why small variations are important There are two categories of variation continuous and discontinuous. It was known in Darwin's day that most variation is continuous. An example of continuous variation is the distribution of heights in people. Humans grade smoothly from one extreme to the other (short to tall), with all the intermediate types (in this case, heights) represented. However, Darwin's contemporaries also knew about discontinuous variation, in which a number of distinct types exist with no intermediates. In humans, height is also subject to discontinuous variation. For example. There is a genetic condition called achondroplasia, which causes affected individuals to be much shorter than other people, have proportionately shorter arms and legs, and bear a variety of other distinctive features. Discontinuous variants are usually quite rare in nature. Nonetheless, many of Darwin's contemporaries, who convinced of the reality of evolution, believed that new species arise as discontinuous variants. * Discontinuous variation is not important for the evolution of complex adaptations because complex adaptations are extremely unlikely to arise in a single jump. Unlike most of his contemporaries, Darwin thought the discontinuous variation did not play an important role in evolution.

Complex adaptations can arise through the accumulation of small random variations by natural selection. Darwin argued that continuous variation is essential for the evolution of complex adaptations.

5.6 Darwin's Difficulties Explaining Variation

Darwin's The Origin of Species was a best seller during his day, but his proposal that new species and other major evolutionary changes arise by the accumulation of small variations through natural selection was not widely embraced. Most educated people accepted the idea that new species arise through the transformation of existing species, and many scientists accepted the idea that natural selection is the most important cause of organic change (although by the turn of the century even this consensus broke down, particularly in the United States). But only a minority endorsed Darwin's view that major changes occur through the accumulation of small variations. Darwin couldn't convinces his contemporaries that evolution occurred through the accumulation of small variations because he couldn't explain how variation is maintained. Darwin's critics raised a telling objection to his theory: the actions of blending and selection would both inevitably deplete variation in populations and make it impossible for natural selection to continue. These were potent objections that Darwin was unable to resolve in his lifetime because he and his contemporaries did not yet understand the mechanics of inheritance.

Figure 5-3 : Darwin around age 71

Everyone could readily observe that many of the characteristics of offspring are an average of the characteristics of their parents. Most people, including Darwin, believed this phenomena to be caused by the action of blending inheritance, a model of inheritance that assumes the mother and father each contribute a hereditary substance that mixes, or —blends,‖ to determine the characteristics of the offspring. Shortly after publication of The Origin of Species, a Scottish engineer named Fleeming Jenkin published a paper in which he clearly showed that with blending inheritance there could be little or no variation available for selection to act on. The following example shows why Jenkin's argument was so compelling. Suppose a population of one species of Darwin's finches display two forms, tall and short. Further suppose that a biologist control mating so that every mating is between a tall individual and a short individual. Then, with blending inheritance all of the offspring will be the same intermediate height, and their offspring will be the same height as they are. All of the variation for height in the population will disappear in a single generation. With random mating, the same thing will occur, though it will take longer. If inheritance were purely a matter of blending parental traits, then Jenkin would have been right about its effect on variation. However, genetics accounts for the fact that offspring are intermediate between their parents and does not assume any kind of blending. Another problem arose because selection works by removing variants from populations. For example, if finches with small beaks are more likely to die than finches with larger beaks over the course of many generations, eventually all that will be left are birds with large beaks. There will be no variation for beak size, and Darwin's second postulate holds that without variation there can be no evolution by natural selection. For example, suppose the environment changes so that individuals with small beaks are less likely to die than those with large are. The average beak size in the population will not decrease because there are no small-beaked individuals. Natural selection destroys the variation required to create adaptations.

Even worse, as Jenkin also pointed out, there was no explanation of how a population might evolve beyond its original range of variation. The cumulative evolution of complex adaptations requires that populations move far outside their original range of variation. Selection can cull away some characters from a population, but how can it lead to new types not present in the original population? This apparent contradiction was a serious impediment to explaining the logic of evolution. How could elephants, moles, bats, and whales all descend from an ancient shrew-like insectivore unless there is some mechanism for creating new variants not present at the beginning? For that matter, how could all the different breeds of dogs have descended from their one common ancestor, the wolf. Remember that Darwin and his contemporaries knew there were two kinds of variation: continuous and discontinuous. Because Darwin believed complex adaptations could arise only through the accumulation of small variations, he thought discontinuous variants were unimportant. However, many biologists thought the discontinuous variants. Called —sports by 19th-century animal breeders, were the key to evolution because they solved the problem of the blending effect. The following hypothetical example illustrates why. Suppose that was slightly more red would have only a small advantage and would be rapidly swamped by blending. In contrast, an all-red bird would have a large enough selective advantage to overcome the effects of blending and could increase its frequency in the population. Darwin's letters show that these criticisms worried him greatly, and although he tried a variety of counter arguments, he never found a satisfactory one. The solution to these problems required an understanding of genetics, which was not available for another half century. As we will see, it was not until well into the 20th century that geneticists came to understand how variation is maintained, and Darwin's theory of evolution was generally accepted.

CHAPTER 6
GENETICS
6.1 Mendelian Genetics

Although none of the main participants in the 20th-century debate about evolution knew it, the key experiments necessary to understand how genetic inheritance really worked had already been performed by an obscure Silesian monk, Gregor Mendel, living in what is now Slovakia. The son of peasant farmers, Mendel was recognized by his teachers as an extremely bright student, and he enrolled in the University of Vienna to study the natural science. While he was there, Mendel received a first-class education from some of the scientific luminaries of Europe. Unfortunately, Mendel had an extremely nervous disposition – every time he was faced with an examination, he became physically ill, taking months to recover. As a result, he was forced to leave the university, after which he joined a monastery in the city of Brno, more or less because he needed a job. Once there, he continued to study inheritance, an interest he had developed in Vienna. Mendel discovered how inheritance worked through careful experiments with plants. Between 1865 and 1873, using the common edible garden pea plant Mendel isolated a number of traits with only two forms, or variants. For example, one of the traits he studied was the color of peas. This trait had two variants, yellow and green. He also studied pea texture, a trait that also had two variants, wrinkled and smooth Mendel cultivated populations of plants in which these traits bred true, meaning that the traits did not change from one generation to the next.

Figure 6-1 :Mendel

For example, crosses (mating) between plants that bore yellow peas consistently produced offspring with yellow peas. Mendel performed a large number of crosses between these different kinds of true-breeding peas. For example, a series of crosses between green and yellow variants yielded offspring that all bore yellow peas matching only one of the parent plants. Mendel's next step was to perform crosses among the offspring of these crosses. Before going further, we need to establish a way to keep track of the results of the matings. Geneticists refer to the original founding population as the Fo generation, the offspring of the original founders as the Fi generation, and so on. In this case, the original true-breeding plants constitute the F0 generation, and the plants created by crossing true-breeding parents constitute the F1 generation. The offspring of the F1 generation will be the F2 generation. When members of the F1 generation (all of whom bore yellow peas) were crossed, some of the offspring produced yellow seeds and some produced green seeds. Unlike most of the people who had experimented with plant crosses before, Mendel did a large number of these kinds of crosses and kept careful count of the numbers of each kind of individual that resulted. These data showed that in the F2 generation there were three individuals with yellow seeds for everyone with green seeds. Mendel was able to formulate two principles that accounted for his experimental results. Mendel derived two insightful conclusions from his experimental results.

1. The observed characteristics of organisms are determined jointly by two particles, one inherited from the mother and one from the father. The American geneticist T. H. Morgan later named these particles genes.

2. Each of these two particles or genes is equally likely to be transmitted when gametes (eggs and sperm) are formed. Modern scientists call this independent assortment. These two principles account for the pattern of results in Mendel's breeding experiments, and as we shall see, they are the key to understanding how variation is preserved.

6.2 Cell Division and the Role of Chromosomes in Inheritance

Nobody paid any attention to Mendel's results for almost 40 years. Mendel thought his findings were important, so he published them in 1866 and sent a copy of the paper to Karl Nageli, a very prominent botanist. Nageli was studying inheritance and should have understood the importance of Mendel's experiments. Instead, Nageli dismissed Mendel's work, perhaps because it contradicted his own results or because Mendel was an obscure monk. Soon after this, Mendel was elected abbot of his monastery and was forced to give up his experiments. His ideas did not resurface until the turn of the 20th century, when several botanists independently replicated Mendel's experiments and rediscovered the laws of inheritance. In 1896, the Dutch botanist Hugo de Vries unknowingly repeated Mendel's experiments with poppies. Instead of publishing his results immediately, he cautiously waited until he had replicated his results with more than 30 plant species.

Then in 1900, just as de Vries was ready to send off a manuscript describing his experiments, a colleague sent him a copy of Mendel's paper. Poor de Vries, his hot new results were already 30 years old! About the same time, two other European botanists, Carl Correns and Erich von Tschermak, also duplicated Mendel's breeding experiments, derived similar conclusions, and again discovered that they had all been done before. Correns and von Tschermak graciously acknowledged Mendel's primacy in discovering the laws of inheritance, but de Vries was less magnanimous. He primacy in discovering the laws of inheritance, but de Vries was less magnanimous. He did not cite Mendel in his treatise on plant genetics and refused to sign a petition advocating the construction of a memorial in Brno commemorating Mendel's achievements. * When Mendel's results were rediscovered, they were widely accepted because scientists now understood the role of chromosomes in the formation of gametes. By the time Mendel's experiments were rediscovered in 1900, it was well known that virtually all living organisms are built out of cells. Moreover, careful al work had shown that all the cells in complex organisms arise from a single cell through the process of cell division. Between the time of Mendel's initial discovery of nature of inheritance and its rediscovery at the turn of the century, a crucial feature of cellular anatomy was discovered – the chromosome. Chromosomes are small, linear bodies contained in every cell and replicated during cell division. Moreover, scientists had also learned that chromosomes are replicated in a special kind of cell division that creates gametes. As we will see in the next section, this research provides a simple material explanation for Mendel's results. Our current model of cell division, which was developed in small steps by a number of different scientists, is summarized in the following sections.

6.3 Mitosis and Meiosis

When plants and animals grow, their cells divide. Every cell contains within it a body called the nucleus; when cells divide, their nuclei also divide. This process of ordinary cell division is called mitosis. As mitosis begins, a cloud of material begins to form in the nucleus, and gradually this cloud condenses into a number of linear chromosomes. The chromosomes can be distinguished under the microscope by their shape and by the way that they stain. (Stains are dyes added to cells in the laboratory that allow researchers to distinguish different parts of a cell). Different organisms have different numbers of chromosomes, but in diploid organisms, chromosomes come in homologous pairs (pairs whose members have similar shapes and staining patterns). All primates are diploid, but other organisms have a variety of arrangements. Diploid organisms also vary in the number of chromosome pairs their cells have. The fruit fly Drosophila has four pairs of chromosomes, humans have 23, and some organisms have many more. Mitosis has two features that suggest that the chromosomes play an important role is determining the properties of organisms. First, mitosis involves duplication of the original set of chromosomes, so that each new daughter cell has an exact copy of the chromosomes present in its parent. This means that as an organism grows and develops through a sequence of mitotic divisions, every cell will have the same chromosomes that were present when the egg and sperm united. Second, the material that makes up the chromosome is present even when cells are not dividing. Cells spend little of their time dividing; most of the time they are in a —resting‖ period, doing what they are supposed to do as liver cells, muscle cells, bone cells, and so on. During the resting period, chromosomes are not visible. However, the material that makes up the chromosomes is always present in the cell. * In meiosis, the special cell-division process that produces gamates, only half of the chromosomes are transmitted from the parent cell to the gamete. * Ordinary cell division, called mitosis, creates two copies of the chromosomes present in the nucleus.

The sequence of events that occur during mitosis is quite different from the sequence of events that occur during meiosis, the special form of cell division leading to the production of gametes. The key feature of meiosis is that each gamete contains only one copy of each chromosome, while cells that undergo mitosis contain a pair of homologous chromosomes. Cells that contain only one copy of each chromosome are said to be haploid. When a new individual is conceived, a haploid sperm from the father unites with a haploid egg from the mother to produce a diploid zygote. The zygote is a single cell that then divides mitotically over and over again to produce the millions and millions of cells that make up an individual's body.

6.4 Chromosomes and Mendel's Experimental Results

In 1902 less than two years after the discovery of Mendel's findings, Walter Sutton, a young graduate student at Columbia University, made the connection between chromosomes and the properties if inheritance discovered by Mendel's principles. Recall that the first of Mendel's two principles states that an organism's observed characteristics are determined by particles acquired from each of the parents. This fits with the idea that genes reside on chromosomes because individuals inherit one copy of each chromosome from each parent. The idea that observed characteristics are determined by genes from both parents is consistent with the observation that mitosis transmits a copy of both the maternal and the paternal chromosomes. Mendel's second principle states that genes segregate independently. The observation that meiosis involves the creation of gametes with only one of the two possible chromosomes from each homologous pair is consistent with to notions: (I) that one gene is inherited from each parent, and (2) that each of these genes is equally likely to be transmitted to gametes. Not everyone agreed with Sutton, but over the next 15 years. T. H. Morgan and his colleagues at Columbia performed many experiments that proved Sutton right. [29-30] * Mendel's two principles can be deduced from the assumption that genes are carried on chromosomes.

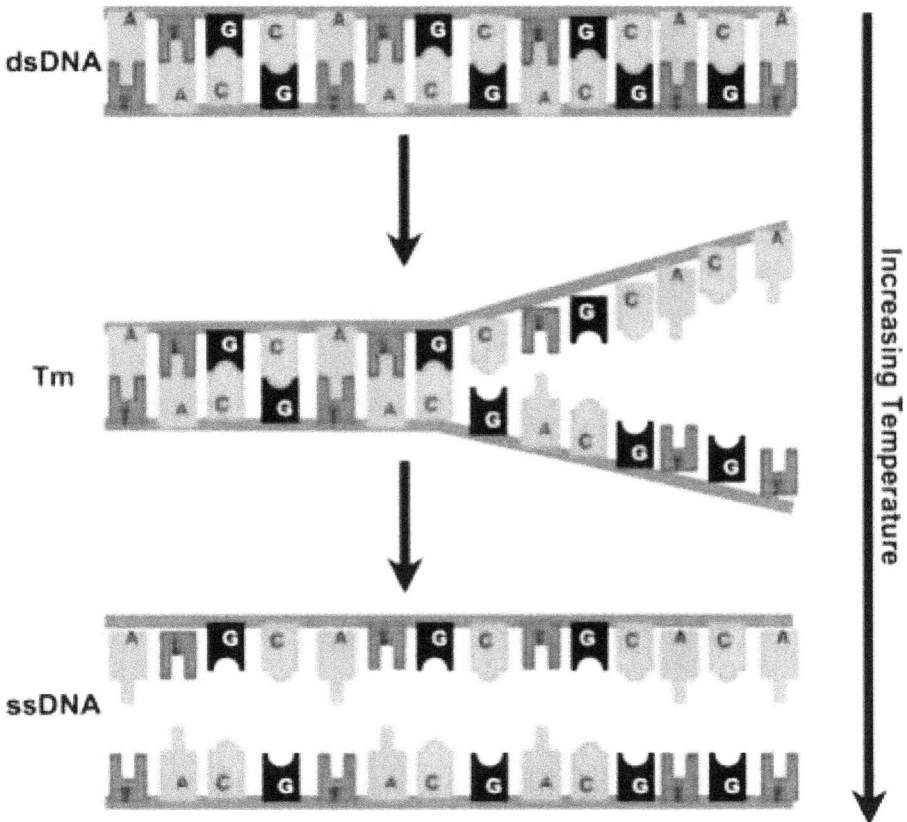

Figure 6-2 : The process of gene expression

6.5 Summary of Mendelian Genetics

Although none of the main participants in the 20th-century debate about evolution knew it, the key experiments necessary to understand how genetic inheritance really worked had already been performed by an obscure Silesian monk, Gregor Mendel, living in what is now Slovakia. The son of peasant farmers, Mendel was recognized by his teachers as an extremely bright student, and he enrolled in the University of Vienna to study the natural science. While he was there, Mendel received a first-class education form some of the scientific luminaries of Europe.

Unfortunately, Mendel had an extremely nervous disposition – every time he was faced with an examination, he became physically ill, taking months to recover. As a result, he was forced to leave the university, after which he joined a monastery in the city of Brno, more or less because he needed a job. Once there, he continued to study inheritance, an interest he had developed in Vienna. Mendel discovered how inheritance worked through careful experiments with plants. Between 1865 and 1863, using the common edible garden pea plant Mendel isolated a number of traits with only two forms, or variants. For example, one of the traits he studied was the color of peas. This trait had two variants, yellow and green. He also studied pea texture, a trait that also had two variants, wrinkled and smooth Mendel cultivated populations of plants in which these traits bred true, meaning that the traits did not change from one generation to the next. For example, crosses (mating) between plants that bore yellow peas consistently produced offspring with yellow peas. Mendel performed a large number of crosses between these different kinds of true-breeding peas. For example, a series of crosses between green and yellow variants yielded offspring that all bore yellow peas matching only one of the parent plants. Mendel's next step was to perform crosses among the offspring of these crosses. Before going further, we need to establish a way to keep track of the results of the matings. Geneticists refer to the original founding population as the F0 generation, the offspring of the original founders as the Fi generation, and so on. In this case, the original true-breeding plants constitute the F0 generation, and the plants created by crossing true-breeding parents constitute the F1 generation. The offspring of the F1 generation will be the F2 generation. When members of the F1 generation (all of whom bore yellow peas) were crossed, some of the offspring produced yellow seeds and some produced green seeds. Unlike most of the people who had experimented with plant crosses before, Mendel did a large number of these kinds of crosses and kept careful count of the numbers of each kind of individual that resulted. These data showed that in the F2 generation there were three individuals with yellow seeds for every one with green seeds.

Mendel was able to formulate two principles that accounted for his experimental results. Mendel derived two insightful conclusions from his experimental results. 1. The observed characteristics of organisms are determined jointly by two particles, one inherited from the mother and one from the father. The American geneticist T. H. Morgan later named these particles genes.

2. Each of these two particles or genes is equally likely to be transmitted when gametes (eggs and sperm) are formed. Modern scientists call this independent assortment. These two principles account for the pattern of results in Mendel's breeding experiments, and as we shall see, they are the key to understanding how variation is preserved.

6.6 Molecular Genetics

In the first half of the 1900s, geneticists made substantial progress in describing the cellular events that took place during meiosis and mitosis, and in understanding the chemistry of reproduction. For instance, by the middle of the 20th century it was known that chromosomes contain two structurally complex molecules – protein and deoxyribonucleic acid, DNA. It had also been determined that the particle of heredity postulated by Mendel was DNA, not protein, though exactly how DNA might contain and convey the information essential to life was still a mystery. But in the early 1950s, two young biologists at Cambridge University, Francis Crick, and James Watson, made a discovery that revolutionized biology: they deduced the structure of DNA. Watson and Crick's elucidation of the structure of DNA was the well-spring of a great flood of research that continues to provide a deep and powerful understanding of how life works at the molecular level. We now know how DNA stores information and how this information controls the chemistry of life, and this knowledge explains why heredity leads to the patterns Mendel described in pea plants, and why there are sometimes new variations.

*Understanding the chemical nature of the gene is critical to the study of human evolution:

(1) molecular genetics links biology to chemistry and physics, and
(2) molecular methods help us to reconstruct the evolutionary history of the human lineage.

* Genes are segments of a long molecule called DNA, which is contained in chromosomes. Modern molecular genetics, the product of Watson and Crick's discovery, is a field of great intellectual excitement producing many new discoveries of practical value in medicine and agriculture. However, progress in molecular genetics has not yet fundamentally changed our understanding if how morphology and behavior evolve. Morphological and behavioral traits are usually affected by many genes that interact in ways that are for the most part still too complex to understand at the molecular level. There are several important

exceptions to this generalization. For example, in recent years there has been rapid progress in our understanding of the genes that control the development of body size and shape, and the genetic basis of certain forms of behavior. Nonetheless, an understanding of molecular genetics is crucially important in understanding the evolution of the human phenotype for two reasons:

1. Molecular genetics links biology to chemistry and physics. One of the grandest goals of science is to provide a single, consistent explanatory framework for the way works. We want to place evolution in this grand scheme of scientific explanation. It is important to be able to explain not only how new species of plants and animals arise, but also the evolution of a wide range of phenomena – form the origin of stars and galaxies of the rise of complex societies. Modern molecular biology is of profound importance because it explains how life and evolution work at the level of physics and chemistry, and chemistry, and it thereby connects physical and geochemical evolution to Darwinian processes.

2. Molecular genetics provides important data for reconstructing evolutionary history. In recent years, molecular genetics have used information about variation in DNA sequences to reconstruct the history of particular lineages. To see how this is done, you have to understand a little about the molecular biology of the gene.

CHAPTER 7

DNA & GENES

7.1 The structure of DNA and Genes

A molecule called deoxyribonucleic acid (DNA), which contains the biological instructions that make each species unique gives rise to every living species on Earth. DNA, along with the instructions it contains, is passed from adult organisms to their offspring during reproduction.

Where is DNA found

DNA is found inside a special area of the cell called the nucleus. Because the cell is very small, and because organisms have many DNA molecules per cell, each DNA molecule must be tightly packaged. This packaged form of the DNA is called a chromosome. During DNA replication, DNA unwinds so it can be copied. At other times in the cell cycle, DNA also unwinds so that its instructions can be used to make proteins and for other biological processes. But during cell division, DNA is in its compact chromosome form to enable transfer to new cells. Researchers refer to DNA found in the cell's nucleus as nuclear DNA. An organism's complete set of nuclear DNA is called its genome. Besides the DNA located in the nucleus, humans and other complex organisms also have a small amount of DNA in cell structures known as mitochondria. Mitochondria generate the energy the cell needs to function properly. In sexual reproduction, organisms inherit half of their nuclear DNA from the male parent and half from the female parent. However, organisms inherit all of their mitochondrial DNA from the female parent. This occurs because only egg cells, and not sperm cells, keep their mitochondria during fertilization.

What is DNA made of

DNA is made of chemical building blocks called nucleotides. These building blocks are made of three parts: a phosphate group, a sugar group and one of four types of nitrogen bases. To form a strand of DNA, nucleotides are linked into chains, with the phosphate and sugar groups alternating.

The four types of nitrogen bases found in nucleotides are: adenine (A),thymine (T), guanine (G) and cytosine (C). The order, or sequence, of these bases determines what biological instructions are contained in a strand of DNA. For example, the sequence ATCGTT might instruct for blue eyes, while ATCGCT might instruct for brown. Each DNA sequence that contains instructions to make a protein is known as a gene. The size of a gene may vary greatly, ranging from about 1,000 bases to 1 million bases in humans. The complete DNA instruction book, or genome, for a human contains about 3 billion bases and about 20,000 genes on 23 pairs of chromosomes.

What does DNA do

DNA contains the instructions needed for an organism to develop, survive, and reproduce. To carry out these functions, DNA sequences must be converted into messages that can be used to produce proteins, which are the complex molecules that do most of the work in our bodies.

How are DNA sequences used to make proteins

DNA's instructions are used to make proteins in a two-step process. First, enzymes read the information in a DNA molecule and transcribe it into an intermediary molecule called messenger ribonucleic acid, or mRNA. Next, the information contained in the mRNA molecule is translated into the "language" of amino acids, which are the building blocks of proteins. This language tells the cell's protein making machinery the precise order in which to link the amino acids to produce a specific protein. This is a major task because there are 20 types of amino acids, which can be placed in many different orders to form a wide variety of proteins.

Who discovered DNA

The German biochemist Frederich Miescher first observed DNA in the late 1800s. But nearly a century passed from that discovery until researchers unraveled the structure of the DNA molecule and realized its central importance to biology. For many years, scientists debated which molecule carried life's biological instructions. Most thought that DNA was too simple a molecule to play such a critical role. Instead, they argued that proteins were more likely to carry out this vital function because of their greater complexity and wider variety of forms. The importance of DNA became clear in 1953 thanks to the work of James Watson, Francis Crick, Maurice Wilkins, and Rosalind Franklin. By studying X-ray diffraction patterns and building models, the scientists figured out the double helix structure of DNA - a structure that enables it to carry biological information from one generation to the next.

What is the DNA Double Helix

Scientists use the term "double helix" to describe DNA's winding, two-stranded chemical structure. This shape - which looks much like a twisted ladder - gives DNA the power to pass along biological instructions with great precision. To understand DNA's double helix from a chemical standpoint, picture the sides of the ladder as strands of alternating sugar and phosphate groups - strands that run in opposite directions. Each "rung" of the ladder is made up of two nitrogen bases, paired together by hydrogen bonds. Because of the highly specific nature of this type of chemical pairing, base A always pairs with base T, and likewise C with G. So, if you know the sequence of the bases on one strand of a DNA double helix, it is a simple matter to figure out the sequence of bases on the other strand.

DNA's unique structure enables the molecule to copy itself during cell division. When a cell prepares to divide, the DNA helix splits down the middle and becomes two single strands. These single strands serve as templates for building two new, double stranded DNA molecules - each a replica of the original DNA molecule. In this process, an A base is added wherever there is a T, a C where there is a G, and so on until all of the bases once again have partners. In addition, when proteins are being made, the double helix unwinds to allow a single strand of DNA to serve as a template. This template strand is then transcribed into mRNA, which is a molecule that conveys vital instructions to the cell's protein-making machinery. The repeating four-base structure of DNA allows the molecule to assume a vast number of distinct forms. Furthermore, the staggering number of DNA molecules that exist in nature are equally stable chemically. DNA is not the only complex molecule with many alternative forms, but other molecules have some forms that are less stable than others. Such molecules would be unsuitable for carrying information because the messages would degrade (become garbled) as the molecules changed toward a more stable form. What makes DNA unusual is that all of its nearly infinite number of forms are equally stable. Each DNA configuration is exactly like a message written in an alphabet with letters that stand for each of the four bases (T for thymine, A for adenine, G for guanine, and C for cytosine). Thus, TCGGTAGTAGTTACGG is one message, while ATCCGGATGCAATCCA is another message. Since the DNA in a single chromosome is millions of bases long, there is room for a nearly infinite variety of messages. In addition to preserving a message faithfully, hereditary material must be replicable. Without the ability to make copies of itself, the genetic message that directs the activities of living cells could not be spread to offspring, and natural selection would be impossible. DNA is replicated within cells by a highly efficient cellular machinery: it first unzips the two strands, and then, with the help of other specialized molecular machinery, it adds complementary bases to each of the strands until two identical sugar and phosphate backbones are built. * The chemical structure of DNA consists of two long backbones made of alternating sugar and phosphate molecules. One of four bases (adenine, guanine, cytosine, or thymine) is attached to each of the sugars. The two strands are connected to each other by hydrogen bonds (dotted lines) between certain pairs of bases. Thymine bonds only to adenine, and guanine bonds only to cytosine.

By acting as catalysts, proteins called enzymes are important in helping the genes of DNA build cells. * When DNA is replicated, the two strands are separated, and two daughter DNA strands are formed. Hereditary material – the DNA in genes – determines what the raw materials are transformed into when used to build cells. To understand how genes build cells, we need to understand the chemical process called catalysis. Consider what a cell might do to a glucose molecule, a simple sugar composed of carbon, oxygen, and hydrogen atoms arranged in a particular way. Suppose we put some glucose in a sterile environment containing oxygen at room temperature. What would happen? The answer depends on how long you would be willing to wait. In the short run not much would happen, but after an extremely long time the carbon and hydrogen would combine with oxygen, and the glucose would be converted to carbon dioxide (CO_2) and water (H_2O).

This change might take a very, very long time, but it would eventually happen. Now, suppose we raised the temperature to 10000F. In this case, sugar would be very rapidly converted to CO_2 and H_2O, quite a bit of heat would be released, and perhaps some light would be produced. In other words, the glucose would catch fire or even explode. What is happening here? At room temperature the reaction that forms CO_2 and H_2O form glucose and oxygen proceeds slowly because there is an intermediate state, a particular conformation of the molecules, that must be reached before the reaction occurs. Achieving the intermediate state requires a threshold amount of energy called the activation energy. At room temperature, only a tiny fraction of the molecules has enough energy (move fast enough) to reach the intermediate state. At higher temperatures, more of the molecules move much faster, so more molecules achieve the activation energy, and the reaction proceeds much more rapidly. Sometimes the addition of another material reduces the activation energy, enabling the reaction to proceed at a rapid rate at lower temperatures, even at room temperature. Such materials are called catalysts, and in biological systems many proteins play this role. Proteins that act as catalysts are called enzymes. * Genes regulate the chemical processes of living cells by determining which enzymes are present to catalyze cellular reactions. The best way to understand how enzymes determine the characteristics of organisms is to think of an organism's biochemical machinery as a branching tree. Glucose serves as a food source for many cells, meaning that it provides energy and a source of raw materials for the construction of cellular structures. Glucose might initially undergo any one of an extremely large number of slow-moving reactions. The presence of particular catalytic enzymes will determine which reactions occur rapidly enough to alter the chemistry of the cell. For example, some enzymes lead to the metabolism of glucose and the release of its stored energy. Thus, enzymes act as switches to determine what chemicals will be present in the cell.

At the end of the first branch there is another node representing all of the reactions that could involve the product(s) of the first branch. Here, one or more enzymes will determine what happens next. This picture has been greatly simplified. Real organisms take in many different kinds of compounds, and each compound is involved in a complicated tangle of branches that biochemists call pathway. Real biochemical pathways are very complex which shows the pathway for the conversion of glucose to energy in animal cells. One set of enzymes causes glucose to be shunted to a pathway that yields energy. A different set enzyme causes the glucose to be shunted to a pathway that binds glucose molecules together to form glycogen, a starch that functions as energy storage. The presence of a third set would lead instead to the synthesis of cellulose, a complex molecule that provides the structural material in plants. Enzymes play roles in virtually all cellular processes – from the replication of DNA and division of cells to the contraction and movement of muscles.

The structure of DNA

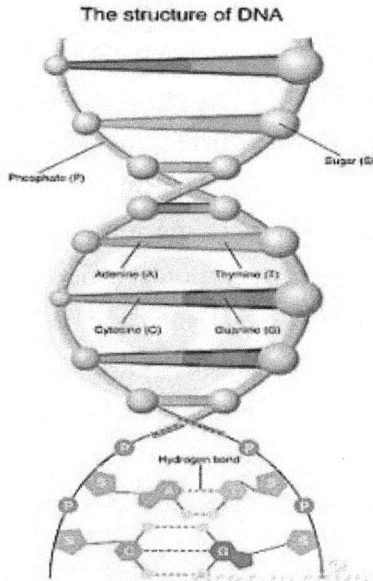

Figure 7-1 : DNA structure

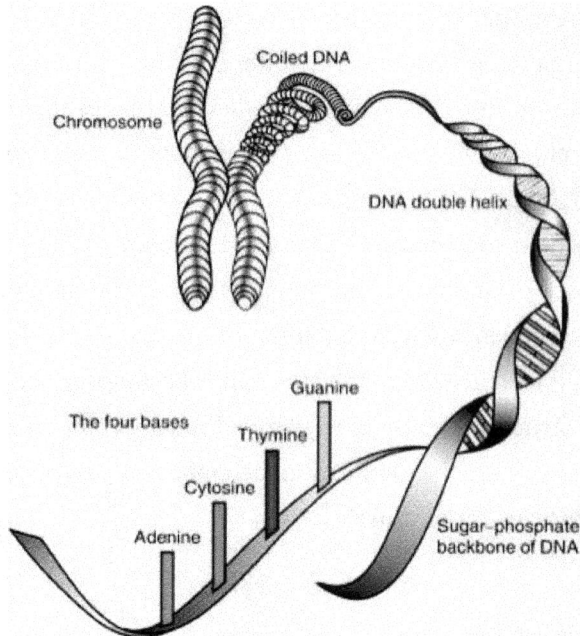

Figure 7-2 : DNA Chromosome components

7.3 DNA Codes for Protein

Once the structure of DNA was known, a new puzzle emerged: how does DNA determine which enzymes will be synthesized? It took more than a decade to discover the amazing answer. To understand this, it is first necessary to know about the structure of enzymes.

* The sequence of amino acids in proteins determines each protein's enzymatic properties.

Like all proteins, enzymes are constructed of amino acids. There are 20 different amino acid molecules. All amino acids have the same chemical backbone, but they differ in the chemical composition of the side chain connected to this backbone (Figure xx). The sequence of amino acid side chains, called the primary structure of the protein, is what makes one protein different from others. You can think of a protein as a very long railroad train in which there are 20 different kinds of cars, each representing a different amino acid. The primary structure is a list of the types of cars in the order that they occur.

When proteins are actually doing their catalytic business, they are folded in complex ways. The three-dimensional shape of the folded protein, called the tertiary structure, is crucial to its catalytic function. The way the protein folds depends on the sequence of amino acid molecules that make up its primary sequence. This means that the function of enzymes depends on the sequence of the amino acids that make them up. (Proteins also have secondary structure, and sometimes also quaternary structure, but to keep things simple, we will ignore them here.)

* DNA specifies the primary structure of protein

Now we return to our original question: how does the information contained in DNA – its sequence of bases – determine the structure of proteins? Remember that DNA encodes messages in a four-letter alphabet. Researchers determined that these letters are combined into three-letter —words‖ called codons, each of which specifies a particular amino acid. Since there are four base times there are

64 possible three-letter combinations for codons (four possibilities for the first base times four for the second times four for the third, or 4 x 4 =64).

Sixty-one of these codons are used to code for the 20 amino acids that make up proteins. For example, the codons GCT, GCC, GCA, and GCG all code for alanine; GAC code for asparagine; and so on. The remaining three codons are —punctuation marks‖ that mean either —start, this is the beginning of the protein‖ or —stop, this is the end of the protein.‖ Thus, if you can identify the base pairs, it is a simple matter to determine what proteins are encoded on the DNA.

*Before DNA is translated into proteins, its message is first transcribed into messenger RNA.

DNA can thought of as a set of instructions for building proteins, but the real work of synthesizing proteins is performed by other molecules. The first step in the translation of DNA into protein occurs when a facsimile of one of the strands of DNA, which will serve as a messenger or chemical intermediary, is made, usually in the cell's nucleus. This copy of ribonucleic acid, or RNA.

RNA is similar to DNA except that it has a slightly different chemical backbone, and the base uracil (denoted U) is substituted for thymine. RNA comes in several forms, all of which aid in protein synthesis. The form of RNA used in this first step is messenger RNA (mRNA). mRNA then leaves the nucleus and migrates to the cytoplasm, the part of the cell outside the nucleus.

* Each kind of amino acid is attached to a transfer RNA molecule that bears the anticodon for that amino acid.

Meanwhile, another essential part of protein synthesis is going on in the cytoplasm. Amino acid molecules are bond to a different kind of RNA, called transfer RNA (tRNA).

These tRNA molecules have a triplet of bases, called an anticodon, at a particular site (Figure xx). Different tRNA molecules have different anticodon sequences and also differ in other ways. Each type of tRNA is bound to the amino acid whose codon binds to the anticodon on the tRNA. For example, one of the codons for the amino acid alanine is the base sequence GCU, and GCU binds only to the anticodon CGA. Thus, the tRNA with the anticodon CGA binds only to the amino acid alanine. Because several different codons correspond to each amino acid, there are more types of tRNA than there are kinds of amino acids.

*The information encoded in DNA determines the structure of proteins in the following way. Inside the nucleus, an mRNA copy is made of the original DNA template. The mRNA is coded in three base codons. Here each codon is given a different color. For example, the sequence AUG codes for the start of a protein and the amino acid methionine, the sequence AGU codes for serine, and the sequence AAA codes for lysine. The mRNA then migrates to the cytoplasm. In the cytoplasm, special enzymes called aminoacyl tRNA synthetases locate a specific kind of tRNA and attach the amino acid whose mRNA codon will bind to the anticodon on the tRNA. For example, the mRNA codon for cysteine is UGG and the appropriate anticodon is ACC because U binds to A and C binds to G. In this diagram, matching mRNA codons and tRNA anticodons are given the same color. The initiation of the process of protein assembly is complicated and involves specialized enzymes.

Once the process is started, each codon of the mRNA binds to the ribosome. Then the matching tRNA is bound to the mRNA, the amino acid is transferred to the growing protein, the tRNA is released, the ribosome shifts to the next codon and the process is repeated.

The tRNAs are bound to the appropriate amino acid by a type of enzyme with the tongue-twisting name aminoacyl-tRNA synthetase. For each type of tRNA there is a distinctive aminoacyl tRNA synthetase that recognizes the tRNA and binds it to the appropriate amino acid. Scientists do not understand exactly how the enzymes accomplish this crucial task; in some case the anticodon sequence seems to be important, while in other cases the enzyme seems to recognize other features of the appropriate tRNA.

* A cellular organelle called the ribosome then synthesizes a particular protein by reading the mRNA copy of the gene.

The next step in the process involves the ribosomes. Ribosomes are small cellular organelles composed of protein and nucleic acid. Organelles are bounded cellular components that perform a particular function, analogous to the way organs like the liver perform a function for the body as a whole. The mRNA first binds to ribosomes at a binding site and then moves through the binding site one codon at a time. As each codon of mRNA enters the binding site, a tRNA with a complementary anticodon is drawn from the complex soup of chemicals inside the cell-bound to the mRNA and added to one end of the growing protein chain. The process repeats for each codon, continuing until the end of the mRNA molecule passes through the ribosome. Voila, a new protein is ready for action.

* Only DNA that is translated into protein is subject to natural selection.

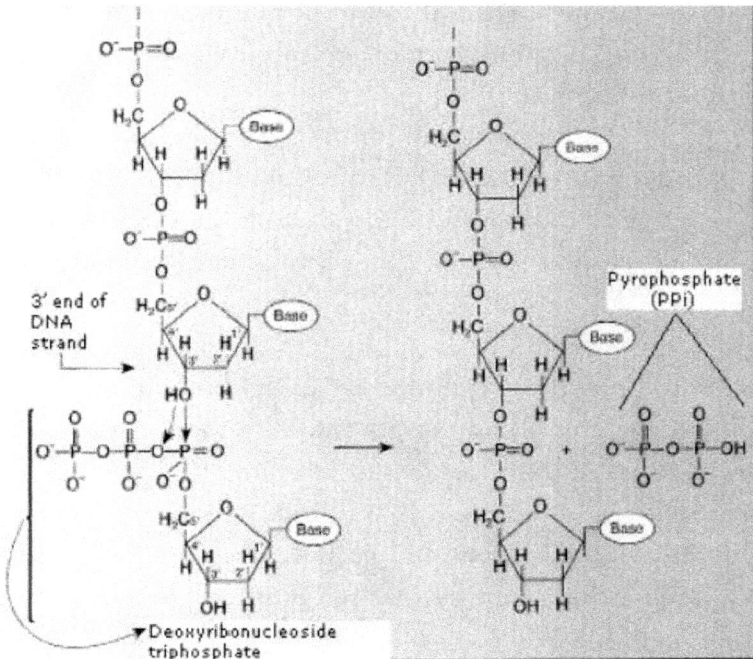

Figure 7-3 : DNA chemical components

7.4 Introns

Introns are not the only kind of DNA that is not translated into proteins. Chromosomes also contain a lot of DNA that is composed of simple repeated patterns. DNA found in introns and in simple repeated sequences is not translated into proteins and doesn't have any direct effect on phenotype. Consequently, this DNA is not subject to natural selection. Fruit flies with the repeated sequence ATATT would be no more likely to survive and reproduce than those with ATAAT. It turns out that the bulk of the DNA in eukaryotes is found in introns and simple repeated sequences. Molecular geneticists estimate that only about 5% of human DNA codes for proteins. Thus, the evolution of 95% of human DNA is not controlled by natural selection, but by random processes like mutation. This fact is very important because it allows us to date distant evolutionary events using genetic information.

In summary, chromosomes contain an enormously long molecule of DNA. Genes are short segments of this DNA. A gene's DNA is, after suitable editing, transcribed into mRNA, which in turn is translated into a protein whose structure is determined by the gene's DNA sequence. Proteins determine the properties of living organisms by selectively catalyzing some chemical reactions and not others, and by forming some of the structural characteristics. Many changes in the morphology or behavior of organisms can be traced back to variations in the proteins and genes that build them. It is important to remember that evolution has a molecule basis. These changes in the genetic constitution of populations that we have briefly explored are grounded in the physical and chemical properties of molecules and genes.

CHAPTER 8

THE VERTEBRATE ANIMAL EVOLUTION

8.1 From Fish to Primates

Vertebrate animals have come a long way since their tiny, translucent ancestors swam the world's seas over 500 million years ago. Here's a roughly chronological list of the major vertebrate animal groups, ranging from fish to amphibians to mammals, with some notable extinct reptile lineages (including archosaurs, dinosaurs, and pterosaurs) in between.

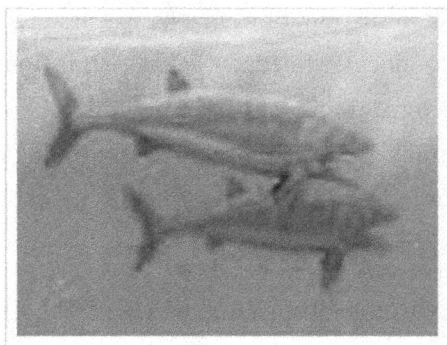

Between 500 and 400 million years ago, life on earth was dominated by prehistoric fish. With their bilaterally symmetric body plans, V-shaped muscles and protected nerve chords running down the lengths of their bodies, these ocean dwellers established the template for later vertebrate evolution (it also didn't hurt that the heads of these fish were distinct from their tails, another surprisingly basic innovation of the Cambrian period). The first prehistoric sharks evolved from their fish forebears about 420 million years ago, and quickly swam to the apex of the undersea food chain.

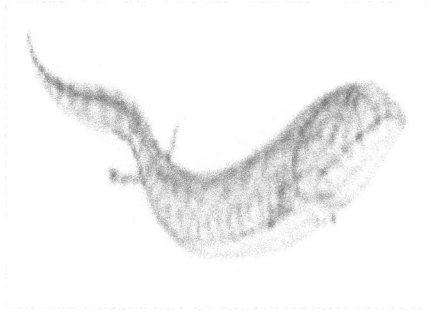

The proverbial "fish out of water," tetrapods were the first vertebrates to climb out of the sea and colonize dry (or at least swampy) land, a key evolutionary transition that occurred somewhere between 400 and 350 million years ago. Crucially, the first tetrapods descended from lobe-finned, rather than ray-finned, fish, which possessed the characteristic skeletal structure that morphed into the fingers, claws, and paws of later vertebrates. (Oddly enough, some of the first tetrapods had seven or eight toes on their hands and feet, and thus counted as evolutionary "dead ends.")

During the Carboniferous period--from about 360 to 300 million years ago--terrestrial life on earth was dominated by prehistoric amphibians. Often considered a mere way station between earlier tetrapods and later reptiles, amphibians were crucially important in their own right, since they were the first vertebrates to figure out a way to colonize dry land (however, these creatures still needed to lay their eggs in water, which severely limited their mobility). Today, amphibians are represented by frogs, toads and salamanders, and their population is rapidly dwindling under environmental stress.

About 320 million years ago--give or take a few million years--the first true reptiles evolved from amphibians (with their scaly skin and semi-permeable eggs, reptiles were free to leave bodies of water behind and venture deep into dry land). The earth's land masses were quickly populated by pelycosaurs, archosaurs (including prehistoric crocodiles), anapsids (including prehistoric turtles), prehistoric snakes, and therapsids (the "mammal-like reptiles" that later evolved into the first mammals); during the late Triassic period, two-legged archosaurs spawned the first dinosaurs, the descendants of which ruled the planet until the end of the Mesozoic Era 175 million years later.

At least some of the first reptiles led partly (or mostly) aquatic lifestyles, but the true age of marine reptiles didn't begin until the appearance of the ichthyosaurs ("fish lizards") during the early to middle Triassic period. The ichthyosaurs overlapped with, and were then succeeded by, long-necked plesiosaurs and muscular pliosaurs, which themselves overlapped with, and were then succeeded by, the exceptionally sleek, vicious mosasaurs of the late Cretaceous period. All of these marine reptiles went extinct 65 million years ago along with their terrestrial dinosaur cousins.

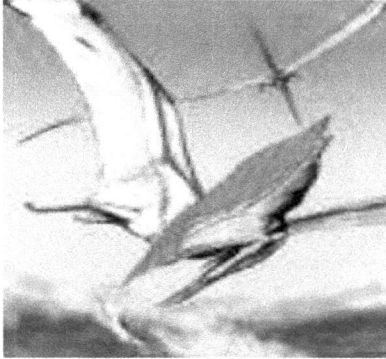

Often mistakenly referred to as dinosaurs, pterosaurs ("winged lizards") were actually a distinct family of reptiles that evolved from archosaurs during the early Triassic period. The pterosaurs of the early Mesozoic Era were fairly small, but some truly gigantic breeds (such as the 200-pound Quetzalcoatlus) dominated the late Cretaceous skies. Like their dinosaur and marine reptile cousins, the pterosaurs went extinct 65 million years ago; contrary to popular belief, they didn't evolve into birds, an honor that belonged to the small, feathered theropod dinosaurs of the Jurassic and Cretaceous periods.

It's difficult to pin down the exact moment when the first true prehistoric birds evolved from their dinosaur forebears; most paleontologists point to the late Jurassic period, about 150 million years ago, on the evidence of distinctly bird-like dinosaurs like Archaeopteryx. However, it's possible that birds evolved multiple times during the Mesozoic Era, most recently from the small, feathered theropods (sometimes called "dino-birds") of the late Cretaceous period. By the way, following the classification system known as "cladistics," it's perfectly legitimate to refer to modern birds as dinosaurs!

109

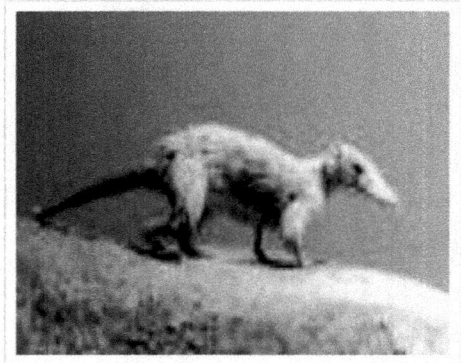

As with most such evolutionary transitions, there wasn't a bright line separating the most advanced therapsids ("mammal-like reptiles") of the late Triassic period from the first true mammals. All we know for sure is that small, furry, warm-blooded, mammallike creatures skittered across the high branches of trees about 230 million years ago and coexisted on unequal terms with dinosaurs right up to the K/T Extinction. Because they were so small and fragile, most Mesozoic mammals are represented in the fossil record only by their teeth, though some species left surprisingly complete skeletons.

After the dinosaurs and marine reptiles vanished off the face of the earth 65 million years ago, the next big theme in vertebrate evolution was the rapid progression of mammals from small, timid, mouse-sized creatures to the giant megafauna of the middle to late Cenozoic Era, including oversized wombats, rhinoceroses, camels, and beavers. Among the mammals that ruled the planet in the absence of dinosaurs and mosasaurs were prehistoric cats, prehistoric dogs, prehistoric elephants, prehistoric horses and prehistoric whales, most species of which went extinct by the end of the Pleistocene epoch (often at the hands of early humans).

Technically, there's no good reason to separate prehistoric primates from the other mammalian megafauna that succeeded the dinosaurs, but it's natural to want to distinguish our human ancestors from the mainstream of vertebrate evolution. The first primates appear in the fossil record as far back as the late Cretaceous period and diversified in the course of the Cenozoic Era into a bewildering array of lemurs, monkeys, apes and anthropoids (the direct ancestors of modern humans). Paleontologists are still trying to sort out the evolutionary relationships of these fossil primates, as new "missing link" species are constantly being discovered.

CHAPTER 9
THE BIOLOGY AND EVOLUTION OF BECOMING
9.1 Becoming Human

One of the hazards of paleoanthropology is the occasional, intense yearning for a time-travel machine. A paleoanthropologist can glean a vast amount of information from a single fossil jaw or kneecap, but so much more remains undiscovered. Some mysteries of human evolution may be solved someday, thanks to an expedition or a brilliant new hypothesis, but some may always hang over us. If a paleoanthropologist could take just a single drive across the African landscape at a crucial moment in hominid history, we might know so much more about where we come from. (Zimmer)

The successive forms of life have been produced by the process of 'descent with modification', as Darwin called it, and are related to each other by a branching genealogy, the tree of life. We human beings are most closely related to chimpanzees and gorillas, with whom we shared a common ancestor 6 to 7 million years ago. (Charlesworth)

The mammals, the group to which we belong, shared a common ancestor with living species of reptiles about 300 million years ago. All vertebrates (mammals, birds, reptiles, amphibia, fishes) trace their ancestry back to a small fish-like creature that lacked a backbone, which lived over 500 million years ago. Further back in time, it becomes increasingly difficult to discern the relationships between the major groups of animals, plants, and microbes, but, as we shall see, there are clear signs in their genetic material of common ancestry. Less than 450 years ago, all European scholars believed that the Earth was the center of a universe of at most a few million miles in extent, and that the planets, Sun, and stars all rotated around this center. Less than 250 years ago, they believed that the universe was created in essentially its present state about 6,000 years ago, although by then the Earth was known to orbit the Sun like other planets, and a much larger size of the universe was widely accepted. (Charlesworth)

Less than 150 years ago, the view that the present state of the Earth is the product of at least tens of millions of years of geological change was prevalent among scientists, but the special creation by God of living species was still the dominant belief. The relentless application of the scientific method of inference from experiment and observation, without reference to religious or governmental authority, has completely transformed our view of our origins and relation to the universe, in less than 500 years. In addition to the intrinsic fascination of the view of the world opened up by science, this has had an enormous impact on philosophy and religion. The findings of science imply that human beings are the product of impersonal forces, and that the habitable world forms a minute part of a universe of immense size and duration. Whatever the religious or philosophical beliefs of individual scientists, the whole program of scientific research is founded on the assumption that the universe can be understood on such a basis.

9.2 The Processes of Evolution

To understand life on Earth, we need to know how animals (including humans), plants, and microbes work, ultimately in terms of the molecular processes that underlie their functioning. This is the 'how' question of biology; an enormous amount of research during the last century has produced spectacular progress towards answering this question. This effort has shown that even the simplest organism capable of independent existence, a bacterial cell, is a machine of great complexity, with thousands of different protein molecules that act in a coordinated fashion to fulfill the functions necessary for the cell to survive, and to divide to produce two daughter cells. This complexity is even greater in higher organisms such as a fly or human being. These start life as a single cell, formed by the fusion of an egg and a sperm. There is then a delicately controlled series of cell divisions, accompanied by the differentiation of the resulting cells into many distinct types. The process of development eventually produces the adult organism, with its highly organized structure made up of different tissues and organs, and its capacity for elaborate behavior.

Our understanding of the molecular mechanisms that underlie this complexity of structure and function is rapidly expanding. Although there are still many unsolved problems, biologists are convinced that even the most complicated features of living creatures, such as human consciousness, reflect the operation of chemical and physical processes that are accessible to scientific analysis.

Biological evolution involves changes over time in the characteristics of populations of living organisms. The timescale and magnitude of such changes vary enormously. Evolution can be studied during a human lifetime, when simple changes occur in a single character, such as the increase in the frequency of strains of bacteria resistant to penicillin within a few years of the widespread medical use of penicillin to control bacterial infections (as discussed in Chapter 5).

At the other extreme, evolution involves events such as the emergence of a major new design of organisms, which may take millions of years and require changes in many different characteristics, as in the transition from reptiles to mammals (see Chapter 4). A key insight of the founders of evolutionary theory, Charles Darwin, and Alfred Russel Wallace, was that changes at all levels are likely to involve the same types of processes. Major evolutionary changes largely reflect changes of the same type as more minor events, accumulated over longer time periods.

Evolutionary change ultimately relies on the appearance of new variant forms of organisms: mutations. Natural selection is the most important of these processes for evolutionary changes that involve the structure, functioning, and behavior of organisms (see Chapter 5). In their papers of 1858, published in the Journal of the Proceedings of the Linnaean Society, Darwin and Wallace laid out their theory of evolution by natural selection with the following argument: Many more individuals of a species are born than can normally live to maturity and breed successfully, so that there is a struggle for existence. There is individual variation in innumerable characteristics of the population, some of which may affect an individual's ability to survive and reproduce. The successful parents of a given generation may therefore differ from the population as a whole.

There is likely to be a hereditary component to much of this variation, so that the characteristics of the offspring of the successful parents will differ from the characteristics of the previous generation, in a similar way to their parents.

If this process continues from generation to generation, there will be a gradual transformation of the population, such that the frequencies of characteristics associated with greater survival ability or reproductive success increase over time. These altered characteristics originated by mutation, but mutations affecting a particular trait arise all the time regardless of whether or not they are favored by selection. Indeed, most mutations either have no effects on the organism, or reduce its ability to survive or reproduce.

It is the process of increase in frequency of variants that improve survival or reproductive success that explains the evolution of adaptive characteristics, since better performance of the individual's body or behavior will generally contribute to greater survival or reproductive success. Such a process of change will be especially likely if a population is exposed to a changed environment, where a somewhat different set of characteristics is favored from those already established by selection. As Darwin wrote in 1858: But let the external conditions of a country alter . . . Now, can it be doubted, from the struggle each individual has to obtain subsistence, that any minute variation in structure, habits, or instincts, adapting that individual better to the new conditions, would tell upon its vigor and health? In the struggle it would have a better chance of surviving; and those of its offspring that inherited the variation, be it ever so slight, would also have a better chance. Yearly more are bred than can survive; the smallest grain in the balance, in the long run, must tell on which death shall fall, and which shall survive. Let this work of selection on the one hand, and death on the other, go on for a thousand generations, who will pretend to affirm that it would produce no effect . . . There is, however, another important mechanism of evolutionary change, which explains how species can also come to differ with respect to traits with little or no influence on the survival or reproductive success of their possessors, and which are therefore not subject to natural selection. This is especially likely to be true of the large category of changes in the genetic material which have little or no effect on the organism's structure or functioning.

If there is selectively neutral variability, so that on average there are no differences in survival or fertility among different individuals, it is still possible for the offspring generation to differ slightly from the parental generation. This is because, in the absence of selection, the genes in the population of offspring are a random sample of the genes present in the parental population.

Real populations are finite in size, and so the constitution of the offspring population will by chance differ somewhat from that of the parents' generation, just as we do not expect exactly five heads and five tails when we toss a coin ten times. This process of random change is called genetic drift. Even the biggest biological populations, such as those of bacteria, are finite, so that genetic drift will always operate. The combined effects of mutation, natural selection, and the random process of genetic drift cause changes in the composition of a population. Over a sufficiently long period of time, these cumulative effects alter the population's genetic makeup, and can thus greatly change the species' characteristics from those of its ancestors. We referred earlier to the diversity of life, reflected in the large number of different species alive today. (A very much larger number have existed over the past history of life, owing to the fact that the ultimate fate of nearly all species is extinction.

The problem of how new species evolve is clearly a crucial one and is dealt with in Chapter 6. The term `species' is hard to define, and it is sometimes difficult to draw a clear line between populations that are members of the same species, and populations that belong to separate species. In thinking about evolution, it makes sense to consider two populations of sexually reproducing organisms as different species if they cannot interbreed with each other, so that their evolutionary fates are totally independent. Thus, human populations living in different parts of the world are unequivocally members of the same species since there are no barriers to interbreeding if migrant individuals arrive from another place. Such migration tends to prevent the genetic makeup of different populations of the same species from diverging very much. In contrast, chimpanzees and humans are clearly separate species, since humans and chimpanzees living in the same area cannot interbreed. As we shall describe later on, humans also differ much more from chimpanzees in the make-up of their genetic material than they do from each other.

The formation of a new species must involve the evolution of barriers to interbreeding between related populations. Once such barriers form, the populations can diverge under mutation, selection, and genetic drift. This process of divergence ultimately leads to the diversity of life. If we understand how barriers to interbreeding evolve, and how populations subsequently diverge, we will understand the origin of species. An enormous amount of biological data falls into place in the light of these ideas about evolution, which have been put on a firm basis by the development of mathematical theories which can be modeled in detail, just as astronomers and physicists model the behavior of stars, planets, molecules, and atoms in order to understand them more completely, and to devise detailed tests of their theories. Before describing the mechanisms of evolution in more detail (but omitting the mathematics), the next two chapters will show how many kinds of biological observations make sense in terms of evolution, in contrast with special creation and its appeal to ad hoc explanations.

9.3 The evidence for evolution similarities and differences between organisms

The theory of evolution accounts for the diversity of life, with all the well-known differences between different species of animals, plants, and microbes, but it also explains their fundamental similarities. These are often evident at the superficial level of externally visible characters but extend to the finest details of the microscopic structure and biochemical function. In the earlier chapters, we discussed the diversity of life later in this book and describe how the theory of evolution can account for new forms appearing from ancestral ones, but here we focus on the unity of living species. In addition, we will introduce many basic biological facts on which later chapters build.

Similarities between different groups of species Similarities between even widely disparate types of organisms exist at every level, from familiar, externally visible resemblances to profound resemblances in life cycles and the structure of the genetic material.

They are plainly detectable even between creatures as different as ourselves and bacteria. These similarities have a natural and straightforward explanation in the idea that organisms are related through an evolutionary process of descent from common ancestors. We ourselves have obvious similarities to apes, as illustrated in Figure 1A, including similarities in internal characters such as our brain structure and organization. There are lesser similarities to monkeys, and even smaller, but still extremely clear, similarities to other mammals, despite all our differences. Mammals have many similarities to other vertebrates, including the basic features of their skeletons, and their digestive, circulatory, and nervous systems. Even more amazing are similarities with creatures such as insects, for example in their segmented body plans, their common need for sleep, the control of their daily rhythms of sleep and waking, and fundamental similarities in how the nerves work in many different kinds of animals, among other features.

An electron micrograph image of a bacterial cell will show its simple structure, with a cell wall and DNA which is enclosed in a nucleus. Sensory receptor proteins, such as olfactory receptors and light receptors, are used in communication between cells and their environment. Chemical and light signals from the outside world are transformed into electrical impulses that travel along the nerves to the brain. All animals that have been studied use largely similar proteins in chemical and light perception. To illustrate the similarities that have been discovered in cells of different organisms, a myosin (motor) protein, similar to proteins in muscle cells, is involved in signaling in flies' eyes and in the ears of humans; one form of deafness is caused by mutations in the gene for this protein. Biochemists have cataloged the enzymes in living organisms into many different kinds, and every known enzyme (many thousands in a complex animal like us) has a number in an international numbering system. Because so many enzymes are found in cells of very wide biochemists have cataloged the enzymes in living organisms into many different kinds, and every known enzyme (many thousands in a complex animal like ourselves) has a number in an international numbering system.

Because so many enzymes are found in cells of a very wide range of organisms, this system categorizes enzymes by the jobs they perform, not the organism they come from. Some, such as digestive enzymes, snip molecules into pieces, others combine molecules together, while others oxidize chemicals (combine them with oxygen), and so on. The means by which energy is generated by cells from food sources is largely the same for all kinds of cells. In this process, there is an energy source (sugars or fats, in the case of our cells, but other compounds, such as hydrogen sulphide, for some bacteria).

A cell takes the initial compound through a series of chemical steps, some of which release energy. Such a metabolic pathway is organized like an assembly line, with a succession of sub-processes. Each sub-process is carried out by its own protein `machine'; these are the enzymes for the different steps in the pathway. The same pathways operate in a wide range of organisms, and modern biology textbooks show the important metabolic pathways without needing to specify the organism. For example, when lizards tire after running, this is caused by the build-up of the chemical lactic acid, just as in our muscles. Cells have pathways to make chemicals of many different kinds, as well as to generate energy from foods. For example, some of our cells make hairs, some make bone, some make pigments, others produce hormones, and so on. The metabolic pathway by which the skin pigment melanin is made is the same in ourselves, in other mammals, in butterflies with black wing pigments, and even in fungi (for instance in black spores), and many of the enzymes involved in this pathway are also used by plants in making lignin, the main chemical constituent of wood. The fundamental similarity of the basic features of metabolic pathways, from bacteria to mammals, is once again readily understandable in terms of evolution.

There are biosynthetic pathways by which melanin and a yellow pigment are synthesized in mammalian melanocyte cells from their amino acid precursor, tyrosine. Each step in the pathway is catalyzed by a different enzyme. Absence of active tyrosinase enzyme results in albino animals. The melanocyte-stimulating hormone receptor determines the relative amounts of black and yellow pigments. Absence of the antagonist to the hormone leads to black pigment synthesis, but presence of the antagonist sets the receptor to `off', leading to yellow pigment formation.

This is how the yellow versus black parts of tabby cat and brown mouse hairs come to be formed. Mutations that make the antagonist non-functional cause darker coloration; however, black animals are not usually the result of this, but simply have the receptor set to 'on' regardless of the hormone level.

Proteins are very large molecules made up of strings of dozens to a few hundreds of amino acid subunits, each joined to a neighboring amino acid, forming a chain. Each amino acid is a quite complex molecule, with individual chemical properties and sizes. Twenty different amino acids are used in the proteins of living organisms; a particular protein, such as the hemoglobin in our red blood cells, has a characteristic set of amino acids in a particular order. Given the correct sequence of amino acids the protein chain folds up into the shape of the working protein. The complex three-dimensional structure of a protein is completely determined by the sequence of amino acids in its constituent chain or chains; in turn, this sequence is completely determined by the sequence of chemical units of the DNA of the gene that produces the protein, as we will soon explain. Studies of the three-dimensional structures of the same enzyme or protein in widely different species show that these are often extremely similar across huge evolutionary distances, such as between bacteria and mammals, even if the sequence of amino acids has changed greatly. An example is the myosin protein that we have already mentioned, which is involved in signaling in flies' eyes and in mammalian ears. Such fundamental similarities mean that, astonishingly, it is often possible to correct a metabolic defect in yeast cells by introducing a plant or animal gene with the same function.

9.4 The Basis of Heredity is Common to All Organisms

The physical basis of inheritance is fundamentally similar in all eukaryote organisms (animals, plants, and fungi). Our understanding of the mechanism of inheritance, that is the control of individuals' many different characteristics by physical entities that we now call genes, first came from work by Gregor Mendel on garden peas, but the same rules of inheritance apply to other plants and to animals, including humans.

The genes that control the production of metabolic enzymes and other proteins (and thus determine individuals' characteristics) are stretches of DNA carried in the chromosomes of each cell (Figures 6 and 7). The discovery that the chromosomes carry the organism's genes in a linear arrangement was first made in the fruit fly, (Drosophila melanogaster) but it is equally true for our own genome. The order of genes on the chromosomes can be rearranged during evolution, but changes are infrequent, so that sets of the same genes in the same order can be found in the human genome and in the chromosomes of other mammals such as cats and dogs. A chromosome is essentially a single very long DNA molecule encoding hundreds or thousands of genes. The DNA of a chromosome is combined with protein molecules that help to package it in neat coils inside the cell nucleus (resembling the devices used for keeping computer cables tidy).

In higher eukaryotes like ourselves, each cell contains one set of chromosomes derived from the mother through the egg nucleus, and another set derived from the father through the sperm nucleus (Figure 6). In humans, there are 23 different chromosomes in a single maternal or paternal set; in Drosophila melanogaster, which is used for much research in genetics, the chromosome number is five (one of which is tiny). The chromosomes carry the information needed to specify the amino acid sequences of an organism's proteins, together with the controlling DNA sequences that determine which proteins will be produced by the organism's cells.

9.5 What is a Gene

What is a gene, and how does it determine the structure of a protein? A gene is a sequence of the four chemicals `letters of the genetic code, in which sets of three adjacent letters (triplets) correspond to each amino acid in the protein for which the gene is responsible. The gene sequence is `translated' into the sequence of a protein chain; there are also triplets marking the end of the amino acid chain. A change in the sequence of a gene causes a mutation. Most such changes will lead to a different amino acid being placed in a protein when it is being made (but, because there are 64 possible triplets of DNA letters, and only 20 amino acids used in proteins, some mutations do not change the protein sequence).

Across the entire range of living organisms, the genetic code differs only very slightly, strongly suggesting that all life on Earth may have a common ancestor. The genetic code was first studied in bacteria and viruses but was soon checked and found to be the same in humans. Almost every possible mutation that this code can generate in the sequence of the human red blood cell protein hemoglobin has been observed, but mutations that are despite the enormous differences in the modes of life of different organisms, ranging from unicellular organisms to bodies composed of billions of cells with highly differentiated tissues, eukaryote cells undergo similar cell division processes. Single-called organisms such as an amoeba or yeast can reproduce simply by division into two daughter cells. A fertilized egg of a multi-cellular organism, produced by the fusion of an egg and a sperm, similarly divides into two daughter cells (Figure 7). Many further rounds of cell divisions then take place to produce the many cells and tissue types that form the body of the adult organism. In a mammal, there are over 300 different types of cells in the adult body. Each type has a characteristic structure and produces a specific array of proteins. The arrangement of these cells into tissues and organs during development requires elaborately controlled networks of interactions between the cells of the developing embryo. Genes are turned on and off to ensure that the right kind of cell is produced in the right place at the right time.

Many signaling processes involved in development and differentiation of particular tissues, such as nerves, are found to be shared by all multi-cellular animals, while land plants use a rather different set, as might be expected from the fact that the fossil record shows that multi-cellular animals and plants have separate evolutionary origins. When a cell divides, the DNA of the chromosomes is first replicated, so that there are two copies of each chromosome. Cell division is a process with tight controls to ensure that the newly copied DNA sequence undergoes —proof-reading" for errors. Cells have enzymes that, using certain properties of the way DNA is replicated, can distinguish new DNA from the old `template' DNA. This enables most errors in copying to be detected and corrected, ensuring that the template has been faithfully copied before the cell is allowed to proceed to the next step, the division of the cell itself.

The machinery of cell division ensures that each daughter cell receives a complete copy of the set of chromosomes that were present in the parent cell. Most prokaryotes' genes (including those of many viruses) are also sequences of DNA that are organized only slightly different from those carried in eukaryote chromosomes. Many bacteria have just one circular DNA molecule as their genetic material. Some viruses, however, such as those responsible for influenza and AIDS, have genes made of RNA. The proof-reading that occurs in DNA replication does not happen when RNA is copied, and so these viruses have extremely high mutation rates and can evolve very rapidly within the host's body. This means that it is difficult to develop vaccines against them. Eukaryotes and prokaryotes differ greatly in their amounts of noncoding DNA.

9.6 The Bacterium

Escherichia coli(a normally harmless species that lives in our intestines) has about 4,300 genes, and the stretches that code for protein sequences form about 86% of this species' DNA. In contrast, less than 2% of the DNA in the human genome codes for protein sequences. Other organisms lie between these extremes. The fruit fly, Drosophila melanogaster, has about 14,000 genes in about 120 million 'letters of DNA, and about 20% of the DNA is made up of coding sequences. The number of different genes in the human genome is still not precisely known. The current best count comes from the sequencing of the complete genome. This allows geneticists to recognize sequences that are probably genes, based on what we know from genes that had previously been studied. It is a difficult task to find these sequences in the huge amount of DNA that makes up the genome of any species, particularly for our own genome, which has a very large DNA content (25 times as much as the fruit fly). The number of genes in humans is about 35,000, much smaller than had been guessed from the number of cell and tissue types with different functions. The number of proteins a human can make is probably considerably larger than this, because this method of counting cannot detect very small genes, or unconventional ones (for example, genes that lie within other genes, which exist in several organisms).

It is not yet known how much of the non-coding DNA is important for the life of the organism. Although much of it is made up of viruses and other parasitic entities that live in chromosomes, some of it has important functions. As we have already mentioned, there are DNA sequences outside genes that can bind proteins controlling which genes in a cell are 'turned on'. The control of gene activity must be much more important in multi-cellular creatures than in bacteria.

In addition to the discovery that widely different organisms have DNA as their genetic material, modern biology has also uncovered profound similarities in the life cycles of eukaryotes, despite their diversity, which ranges from unicellular fungi such as yeasts, to annual plants and animals, to long-lived (though not immortal) creatures like ourselves and many trees.

Many, though not all, eukaryotes have a sexual stage in each generation, in which the maternal and paternal genomes of the uniting egg and sperm (each made up of a set of some number n of different chromosomes, characteristic of the species in question) combine to make an individual with 2n chromosomes. When an animal makes new eggs or sperm, the n condition is restored by a special kind of cell division. Here, each pair of maternal and paternal chromosomes lines up, and (after exchanging material to form chromosomes that are patchworks partly of paternal and partly of maternal DNA) the chromosome pairs separate from each other in a similar way to the separation of newly replicated chromosomes in other cell divisions. At the end of the process, the number of chromosomes in each egg or sperm cell nucleus is therefore halved, but each egg or sperm has one complete set of the organism's genes. The double set will be restored on the union of egg and sperm nuclei at fertilization.

> *The basic features of sexual reproduction must have evolved long before the evolution of multi-cellular animals and plants, which are latecomers on the evolutionary scene.*

This is clear from the common features displayed in the reproduction of sexual unicellular and multi-cellular organisms, and the similar genes and proteins that have been discovered to be involved in the control of cell division and chromosome behavior in groups as distant as yeast and mammals.

In most single-celled eukaryotes, the 2n cell produced by fusion of a pair of cells, each with n chromosomes, divides immediately to produce cells with n chromosomes, just as described above for germ cell production in multi-cellular animals. In plants, the reduction of chromosome number from 2n to n happens before egg and sperm formation, but the same kind of special cell division is again involved; in mosses, for instance, there is a prolonged life-cycle stage with chromosome number n that forms the moss plant, on which the small 2n parasitic stage develops after eggs and sperm are made and fertilization has occurred.

The complications of such sexual processes are absent from some multi-cellular organisms. In such `asexual' species, mothers produce daughters without a reduction of chromosome number from 2n during egg production. Nevertheless, all multi-cellular asexual organisms show clear signs of being descended from sexual ancestors. For example, common dandelions are asexual; their seeds form without the need for pollen to be brought to the flowers, as is required for most plants to reproduce. This is an advantage to a weedy species like the common dandelion, which speedily generates large numbers of seeds, as anyone who has a lawn can see for themselves. Other dandelion species reproduce by normal mating between individuals, and common dandelions are so closely related to these that they still make pollen that can fertilize the flowers of the sexual species.

9.7 Mutations and Their Effects

Despite the proof-reading mechanisms that correct errors when DNA is copied during cell division, mistakes do occur, and these are the source of mutations. If a mutation results in a change in the amino acid sequence of a protein, the protein may malfunction; for example, it may not fold up correctly and so may be unable to do its job properly. If it is an enzyme, this can cause the metabolic pathway to which it belongs to run slowly, or not at all, as in the case of the albino mutations already mentioned. Mutations in structural or communication proteins may impair cell functions or the organism's development. Many diseases in humans are caused by such mutations.

For instance, mutations in genes involved in controlling cell division increase the risk of cancer developing. As already mentioned, cells have exquisite control systems to ensure that they divide only when everything is in order (proof-reading for mutations must be complete, the cell must show no signs of infection or other damage, and so on). Mutations affecting these control systems can result in uncontrolled cell division, and malignant growth of the cell lineage. Luckily, it is unusual for both members of a pair of genes in a cell to be mutant, and one nonmutant member of the pair is often enough for correct cell functioning. A cell lineage also usually requires other adaptations to become a successful cancer, so malignancy is uncommon. (A blood supply is needed for tumors, and the cells' abnormal characteristics must evade detection by the body.) Nevertheless, understanding cell division and its control is a major part of cancer research. The process is so similar in cells of different eukaryote organisms that the 2001 Nobel prizes in medicine were given for research on cell division in yeast, which showed that a gene involved in the control system of yeast cells is mutated in some human familial cancers.

Mutations that give a predisposition to cancer are rare, as are most other disease-causing mutations. The most common genetic disorder in northern European human populations is cystic fibrosis, but even in this case the non-mutant sequence of the gene involved represents more than 98% of copies of the gene in the population.

Mutations that cause failure of an important enzyme or protein may lower the survival or fertility of affected individuals. The gene sequence that leads to the non-functional enzyme will thus be under-represented in the next generation and will eventually be eliminated from the population. A major role of natural selection is to keep the proteins and other enzymes of most individuals working well.

One important type of mutation leads to a protein not being produced in sufficient amounts by its gene. This could happen because of a problem in the normal control system for that gene, which either does not switch it on when it should do so, does not produce in the right quantities, or stops production of the protein before it is finished.

Other mutations may not abolish an enzyme's production, but the enzyme may be faulty, just as a production line can be hindered or stopped if one of the necessary tools or machines is defective in some way. If one or more of the component amino acids are missing, the protein may not function correctly, and the same can happen if a different amino acid appears at a particular position in the chain, even if all the rest are correct. Mutations causing loss of function can contribute to evolution when selection no longer acts to eliminate them and selectively neutral mutations can spread. About 65% of human olfactory receptor genes are 'vestigial genes' that do not produce working receptor proteins, so we have many fewer olfactory functions than mice or dogs (not surprisingly, given the importance of smell in their daily lives and social interactions, compared with its minor role in ours).

9.8 Biological classification, DNA, and protein sequences

A new and important set of data providing clear evidence that organisms are related to one another through evolution comes from the letters in their DNA, which can now be "read" by the chemical procedure of DNA sequencing. Systems of biological classification based on visible characteristics, which were developed over the past three centuries of study of plants and animals, are now supported by recent work comparing DNA and protein sequences among different species. Measuring the similarity of DNA sequences makes it possible to have an objective concept of relationship among species. For the moment we need only understand that the DNA sequences of a given gene will be most similar for more closely related species, while those of more distantly related species are more different. The amount of difference increases roughly proportionally to the amount of time separating two sequences being compared. This property of molecular evolution allows evolutionary biologists to estimate times of events that cannot be studied in fossils, using a molecular clock. For instance, we have already mentioned changes in the order of an organism's genes on its chromosomes.

A molecular clock can be used to estimate the rate of such chromosomal rearrangements. Consistent with the evolutionary viewpoint, species that we believe to be close relatives, such as humans and rhesus monkeys, have chromosomes that differ by fewer rearrangements than humans and New World primates such as the woolly monkey.

131

9.9 The evidence for evolution: patterns in time and space The fossil record

The fossil record is our only direct source of information on the history of life. To interpret it correctly, it is necessary to understand how fossils are formed, and how scientists study them. When a plant, animal, or microbe dies, the soft parts are almost certain to decay rapidly. Only in unusual environments, such as the arid atmosphere of a desert or the preservative chemicals of a piece of amber, are the microbes responsible for decay unable to break down the soft parts.

Remarkable cases of preservation of soft parts, sometimes going back tens of millions of years in the case of insects trapped in amber, have been found, but these are the exception rather than the rule. Even skeletal structures, such as the tough chitin which covers the bodies of insects and spiders, or the bones and teeth of vertebrates, eventually decay. Their slower rate of disappearance offers, however, an opportunity for minerals to infiltrate them, and eventually replace the original material with a mineralized replica (occasionally this happens to soft parts as well). Alternatively, they may create a mold of their shape as minerals are deposited around them. Fossilization is most likely to happen in aquatic environments, where the deposition of sediment and precipitation of minerals occur at the bottom of seas, lakes, and river estuaries. Remains that sink to the bottom can then turn into fossils, although the chance that this happens for a given individual is extremely small. The fossil record is therefore very biased: marine organisms living in shallow seas, where sediments are continuously formed, have the best fossil record, and flying creatures have the worst.

In addition, the deposition of sediments may be interrupted, for example by a change in climate or by uplift of the seabed. For many types of creatures, we have almost no fossil record; for others, the record is interrupted many times the gaps in the fossil record mean that it is rare to have a long-continued series of remains showing the more or less continuous changes which are expected under the hypothesis of evolution.

In most cases, new groups of animals or plants make their first appearance in the fossil record without any obvious links to earlier forms. The most famous example is the `Cambrian explosion', which refers to the fact that most of the major groups of animals appear for the first time as fossils in the Cambrian period, between 550 and 500 million years ago. Nevertheless, as Darwin argued eloquently in The Origin of Species, the general features of the fossil record provide strong evidence for evolution. The discoveries of paleontologists since his day have reinforced his arguments. In the first place, many examples of intermediate forms have been discovered, connecting groups that were formerly thought to be separated by unbridgeable gaps. The fossil bird-reptile Archaeopteryx, discovered shortly after the publication of The Origin of Species, is perhaps the most famous of these. Archaeopteryx fossils are rare (only six specimens exist). They come from Jurassic limestone from about 120 million years ago that was laid down in a large lake in Germany. These creatures show a mosaic of characteristics, some resembling those of modern birds, such as wings and feathers, and others like those of reptiles, such as a toothed jaw (instead of a beak) and a long tail. Many details of their skeletons are indistinguishable from those of a contemporary group of dinosaurs, but Archaeopteryx differs from them, as it could clearly fly. Other fossils linking birds and dinosaurs have subsequently been found, and it has recently been shown that dinosaurs with feathers existed before.

9.10 Archaeopteryx

Other important intermediates include fossil mammals from the Eocene (about 60 million years ago), with forelimbs and reduced hind limbs adapted to swimming. These link modern whales to animals that belong to the group of cloven-hoofed herbivores that includes cows and sheep. Humans are an excellent example of gaps in the record being filled as more research is done. No fossil remains connecting apes and humans were known at the time of publication of Darwin's 1871 book on human evolution, The Descent of Man. Darwin argued on the basis of anatomical similarities that humans were most closely related to gorillas and chimpanzees and had therefore probably originated in Africa from an ancestor that also gave rise to these apes. A whole series of fossil remains have since been found and accurately dated by the methods described earlier, and new fossils continue to be found. The nearer in time to the present, the more similar are the fossils to modern humans (Figure 10); the earliest fossils that can be assigned clearly to Homo sapiens date from only a few hundred thousand years ago. In agreement with Darwin's inferences, early human evolution probably took place in Africa, and it seems likely that our relatives first entered Eurasia about 1.5 million years ago.

Skulls of some human ancestors and relatives such as a Female gorilla, fossils of two different species of one of the earliest human relatives, Australopithecus, from about 3 million years ago and a fossil of an intermediate between Australopithecus and modern humans called Homo erectus, from about 1.5 million years ago. Also, a fossil Neanderthal human, Homo neanderthalensis, from about 70,000 years ago and finally a Modern human, Homosapien was found.

There are also cases of almost completely continuous temporal sequences of fossils, in which it seems certain that we have a record of change in a single evolving lineage. The best examples come from studies of the results of drilling down into deposits at the bottom of the sea, from which long rock columns can be recovered. This allows very fine-scaled time separation between successive samples of the microorganisms whose innumerable fossilized skeletons form the body of the rock.

Careful measurements of the shapes of the skeletons of creatures such as foraminiferans, which are single-celled marine animals, allows characterization both of the averages and levels of variability of successive populations over a long period of time (Figure 11). Even if there were no graded intermediates in the fossil record, the general features of the record are barely comprehensible except in the light of evolution. Although the fossil record before the Cambrian era is fragmentary, there is evidence for the remains of bacteria and related single-celled organisms going back more than 3.5 billion years. Much later on, there are remains of more advanced (eukaryote) cells, but still no evidence for multi-cellular organisms. Organisms made up of simple clusters of cells appear only about 800 million years ago (MYA), at a time of environmental crisis when the Earth was largely covered with ice. About 700-550 MYA, there is evidence for soft-bodied, multi-cellular animal life.

During the next division of the geological record, the Carboniferous (360-280 MYA), land life-forms become abundant and diverse. The coal deposits, which give this period its name, are the fossilized remains of tree-like plants that grew in tropical swamps, but these are similar to contemporary horsetails and ferns and are unrelated to modern conifers or deciduous trees. Remains of primitive reptiles, the first vertebrates to become fully independent of water, are found at the end of the Carboniferous. In the Permian (280-250 MYA), there is a great diversification of reptiles; some of these have anatomical features that increasingly come to resemble those of mammals (the mammal-like reptiles). Some of the modern groups of insects, such as bugs and beetles, appear. The Permian ends with the largest set of extinctions seen in the fossil record, in which some previously dominant groups such as trilobites disappear completely, and many other groups are nearly wiped out. In the recovery that follows, a variety of new forms appear, both on the land and in the sea. Plants similar to modern conifers and cycads appear in the Triassic(250-200 MYA). Dinosaurs, turtles, and primitive crocodiles appear; right at the end, the first true mammals are found.

The fossil record thus suggests that life originated in the sea over 3 billion years ago, and that for more than a billion years only single-celled organisms related to bacteria existed. This is exactly what is expected on an evolutionary model; the evolution of the machinery needed to translate the genetic code into protein sequences, and the complex organization of even the simplest cell, must have required many steps, the details of which almost defy our imagination. The late appearance in the record of clear evidence for eukaryote cells, with their substantially more complex organization compared with prokaryotes, is also consistent with evolution. The same applies to multi-cellular organisms, whose development from a single cell requires elaborate signaling mechanisms to control growth and differentiation: these could not have evolved before single-celled forms existed. Once simple multi-cellular forms evolved, it is understandable that they rapidly diversified into numerous forms, adapted to different modes of life, as occurred in the Cambrian. The fact that life was exclusively marine for an immense period is also understandable from an evolutionary perspective. Early in the Earth's history, the geological evidence shows that there was very little oxygen in the atmosphere. The consequent lack of protection from ultra-violet radiation by atmospheric ozone, which is formed from oxygen, would have prohibited life on land or even in fresh water. Once sufficient oxygen had built up as a result of the photosynthetic activities of early bacteria and algae, this barrier was removed, and the possibility of the invasion of the land opened up. There is evidence for an increase in atmospheric oxygen levels during the period leading up to the Cambrian, which may have permitted the evolution of larger and more complex animals. Similarly, the appearance of fossils of flying insects and vertebrates after the emergence of life on land makes sense, since it is unlikely that true flying animals could evolve from purely aquatic forms.

The recurrent phenomenon of the emergence of abundant and diverse forms of life, followed by their wholesale extinction (as with the trilobites and dinosaurs) or their reduction to just one or a few surviving forms (like the coelacanths) also makes sense in terms of evolution, whose mechanisms have no foresight and cannot guarantee that their products can survive sudden large environmental changes. Similarly, the rapid diversification of groups after the colonization of a new habitat (as in the invasion of the land), or after the extinction of a dominant rival group (as with the mammals after the disappearance of the dinosaurs), is expected on evolutionary principles. The interpretation of the fossil record in terms of biological knowledge therefore follows the same principle of uniformitarianism that is applied by geologists to the history of the structure of the Earth. The fossil evidence might have shown patterns that falsify evolution. The great evolutionist and geneticist J. B. S. Haldane is alleged to have answered the question of what observation would cause him to abandon his belief in evolution by saying: 'A pre-Cambrian rabbit'. So far, no such fossil has been found.

9.11 Adaptation and Natural Selection

The problem of adaptation is an important task of the theory of evolution to account for the diversity of living organisms within the hierarchical organization of resemblances between them. In an earlier chapter, we emphasized resemblances between different groups, and how they make sense in terms of Darwin's theory of descent with modification.

The second essential part of evolutionary theory is to provide a scientific explanation for the 'adaptation' of living organisms: their appearance of good engineering design, and their diversity in relation to their different ways of living. This will require the longest chapter in this book. There are innumerable remarkable examples of adaptations, and we will just mention a few to illustrate the nature of the problem. The diversity of different kinds of eyes alone is astonishing, and yet makes sense in relation to the environments in which different animals live.

Eyes for seeing underwater are different from those for seeing in air, and the eyes of predators have special adaptations to break the camouflage of prey that have evolved to be difficult to see. Many underwater predators that eat transparent marine animals have eyes with special contrast-increasing systems, including ultra-violet vision and polarized light vision.

9.12 Adaptation and Natural Selection

The problem of adaptation An important task of the theory of evolution is to account for the diversity of living organisms within the hierarchical organization of resemblances between them. In an earlier chapter, we emphasized resemblances between different groups, and how they make sense in terms of Darwin's theory of descent with modification. The second essential part of evolutionary theory is to provide a scientific explanation for the 'adaptation' of living organisms: their appearance of good engineering design, and their diversity in relation to their different ways of living. This will require the longest chapter in this book. There are innumerable remarkable examples of adaptations, and we will just mention a few to illustrate the nature of the problem. The diversity of different kinds of eyes alone is astonishing, and yet makes sense in relation to the environments in which different animals live. Eyes for seeing underwater are different from those for seeing in air, and the eyes of predators have special adaptations to break the camouflage of prey that have evolved to be difficult to see. Many underwater predators that eat transparent marine animals have eyes with special contrast-increasing systems, including ultra-violet vision and polarized light vision. Other before Darwin and Wallace, such adaptations appeared to require a Creator. There seemed no other way to account for the astonishing detail and apparent perfection of many aspects of living organisms, just as the complexity of a watch could not be a purely natural production. The absence of any other explanation was the main support for the Argument from Design developed by 18th-century theologians to prove‘ the existence of a Creator, and the term adaptation was introduced to describe the observation that living things have structures that seem to be useful to them.

> There is no doubt that animals and plants differ from other naturally produced things, such as rocks and minerals, as we acknowledge in the game —animal, vegetable, or mineral.

But the argument from Design overlooks the possibility that there could be natural processes, in addition to those that produce minerals and rocks, mountains and rivers, which can account for living creatures as complex natural productions, without the need for a Designer. The biological explanation of the origin of adaptation replaces the idea of a Designer and is central to post Darwinian evolutionary biology. In this chapter, we describe the modern theory of adaptation and its biological causes and basis. This is based on the theory of natural selection, which we will outline in Chapter 6.

9.13 Artificial Selection and Heritable Variation

A first, very pertinent observation, strongly emphasized by Darwin, is that the modification of organisms by humans is regularly possible and can produce the same appearance of design that we see in nature. This is routinely achieved by artificial selection, or selective breeding from animals and plants with desirable characters. Very striking changes can be produced over a timeframe that is short on the scale of the fossil record of evolution. For example, we have developed many different strains of cabbages, including strange ones like the cauliflower and broccoli, which are mutants that cause monstrous flowers forming a large head, and ones like the brussels sprout in which leaf development is abnormal. Similarly, many breeds of dogs have been bred by humans, with differences very like those observed between different species in nature, as Darwin pointed out.

However, although all Canis species (including coyotes and jackals) are close relatives and can interbreed, dog breeds are not domestications of different wild dog species but have been produced over the past few thousand years (several hundred dog generations) by artificial selection from a single common ancestral species, the wolf. The DNA sequences of dog genes are essentially a subset of wolf sequences, but coyotes (whose ancestor is believed from fossils to have separated from wolves' ancestors a million years ago) are about three times as different from either dogs or wolves as the most different dog/wolf comparison. The differences among dogs in their sequences of the same gene, differences which presumably developed after dogs separated from wolves, can be used to tell how long ago that separation happened (see Chapter 3). The conclusion is that dogs separated from wolves much longer ago than 14,000 years, the date suggested by archaeological records, but not more than 135,000 years ago.

Some of the diversity like in mammals such as the differences in the sizes and shapes of two breeds of dog. The success of artificial selection is possible because heritable variation exists within populations and species (the slight differences between normal individuals. Even without any understanding of inheritance, people have bred from animals and plants that had characteristics they liked or found useful, and over enough generations this process has generated strains of animal and plant species that differ greatly from one another, and from the ancestral forms that were originally domesticated. This shows clearly that individuals within domesticated species must have been different from one another, and that many differences can be passed from parents to their offspring, that is they are heritable. If differences were merely due to the way the animals or plants were treated, selective breeding and artificial selection would have no effect on the next generation. Unless some of the differences were heritable, the breed could improve only by improved husbandry. Every imaginable kind of character can vary heritably. The different breeds of dog differ, as is well known, not just in appearance and size, but also in mental traits such as character and disposition, some tending to be friendly, while others are fierce and suitable as guard dogs. They differ in their interest in scents, and in their inclination to fetch and carry or to swim, and in intelligence.

They differ in the diseases to which they are susceptible, as in the well-known case of Dalmatians being prone to gout. They even differ in the ageing process, with some breeds, such as the Chihuahua, having surprising longevity (their lifespan is almost as long as that of cats), while others, such as the Great Dane, live only about half as long. Although all these characteristics are, of course, affected by environmental circumstances such as good care and treatment, they are strongly influenced by heredity. Similar heritable differences are known in many other domesticated species. To take another example, the qualities of different apple varieties are heritable differences. They include adaptations to different human needs such as early or late harvesting, suitability for cooking or eating, and to the differing climates of different countries. Just as in the case of dogs, other evolutionary processes have gone on in apples at the same time as human selection, and perfection is never attained for all desirable traits. For instance, Coxes are a particularly flavorsome apple, but are highly susceptible to disease.

9.14 Kinds of Heritable Variation

The success of artificial selection is very strong evidence that many kinds of character differences in animals and plants are heritable. There are also many genetic studies showing heritable variation for the characteristics of a wide range of organisms in nature, including many different species of animals, plants, fungi, bacteria, and viruses. Variation originates by well-understood processes of random mutation in the DNA sequences of genes, similar to those that produce human genetic disorders. Most mutations are probably deleterious, like the genetic disorders of humans and farm animals, but advantageous mutations do sometimes occur. Such mutations have led to the resistance of animals to disease (such as the evolution of myxomatosis resistance in rabbits). They are also responsible for a major problem today, pests evolving resistance to chemicals (including resistance of rats to warfarin, or of worms parasitic in humans and farm animals to anti-helminthic chemicals, insecticide resistance in mosquitoes, and antibiotic resistance in bacteria).

Because of their importance to human or animal welfare, many cases are understood in great detail. Heritable differences are also well known in humans. Variation may take the form of `discrete' character differences, such as eye and hair color, as already mentioned. These are variants controlled by differences in single genes, and unaffected by environmental circumstances (or altered only slightly, for instance when a fair-haired person's hair is bleached by the sun).

Common variants like these are called polymorphisms. Conditions such as color blindness are also simple genetic differences but are much rarer variants in human populations. Even behavioral characters may be heritable. Whether fire ant colonies have single or multiple queens seems to be controlled by a difference in a single gene for a protein that binds a chemical involved in recognition of other individuals.

`Continuous' variation is also very evident for many characters in populations, for example the gradations of height and weight among people. This kind of variation is often markedly affected by environmental conditions. The increasing height of successive generations during the 20th century, seen in many different countries, is not due to genetic changes but to changed conditions of life, including better nutrition and fewer serious illnesses during childhood. Nevertheless, there is also some degree of genetic determination for such characters in human populations. This is known from studies of identical and non-identical twins. Non-identical twins are ordinary siblings that happen to be conceived at the same time, and they differ as much as any siblings, but identical twins come from a single fertilized egg that splits into two embryos and are genetically identical. Greater resemblances between identical than non-identical twins have been documented for many characteristics, which must be due to their genetic similarity (care must, of course, be taken that the identical twins are not treated more alike than non-identical pairs - for instance, genetic differences boil down to differences in the 'letters' in the DNA. These often leave the amino acid sequences of proteins unchanged. When the DNA sequences of the same gene are compared between different individuals, differences are seen, though usually fewer than when sequences are compared between different species. For example, copies of the gene for glucose-6- phosphate dehydrogenase, as mentioned in an earlier chapter, one from each of a set of individuals, might be compared.

There may be no differences (so no diversity). If some individuals in the population have a variant sequence of the gene, the difference will show up in some of the comparisons. This is called molecular polymorphism. Geneticists measure such diversity by the fraction of the letters in the DNA sequence that vary between individuals in the population. In the human species, it is usually found that fewer than 0.1% of the DNA letters differ when we compare the same gene's sequence between different people (compared with generally around 1% of the letters being different when a gene's sequences are compared between a human and a chimpanzee).

Variation is higher in some genes and lower in others, and, as one might expect, variation is generally higher in the presumably less important regions of the genome that do not code for proteins than in the coding parts of genes. Humans are rather lacking in variability compared with most other species. For example, DNA polymorphism is much more common in maize (more than 2% of its DNA letters are variable). The distribution of variability within a species can give us useful information. When dogs are bred for different characteristics, breeds are developed that are rather uniform in their characteristics. This is due to strict pedigree rules, which control mating and forbid 'gene flow' between breeds. A characteristic that is desired in one breed, such as fetching, is thus well developed in that breed only, and separate breeds tend to diverge from each other. This isolation between breeds is unnatural, and dogs of different breeds will happily mate and produce healthy young. Much of the variability of dogs is accordingly between breeds. Many natural species live in different, geographically separated populations, and, as one might expect, the amount of diversity in such species as a whole is greater than within a single population, because there are differences between populations.

9.15 Natural Selection and Fitness

A fundamental idea in the theory of evolution under natural conditions is that some heritable character differences affect survival and reproduction. For instance, just as racehorses have been selected for speed (by breeding from winners and their relatives), so antelopes have been naturally selected for speed because the individuals that breed and contribute to the future of their species are those that did not get eaten by predators. Darwin and Wallace realized that this kind of process could explain adaptation to natural conditions. Our ability to modify animals and plants by artificial selection depends on this characteristic having a heritable basis. Provided that there are heritable differences, successful individuals in the wild will likewise pass their genes (and thus often their good characteristics) to their offspring, which will, in turn, possess the adaptive characters, such as speed. For brevity, and to allow one to think in general terms, the word fitness is often used in biological writing to stand for overall ability to survive and reproduce, without the need to specify which characters are involved (just as we use the term 'intelligence' to mean a variety of different abilities). Many different aspects of organisms contribute to fitness. For instance, speed is just one feature affecting antelope fitness. Alertness and the ability to detect predators are also important. Mere survival is not enough, however, and reproductive abilities, such as provisioning and care of the young, are also important for fitness in animals, and the ability to attract pollinators is critical for fitness in flowering plants. The word fitness can accordingly be used to describe selection acting on a wide range of different traits. As with —intelligence,‖ the generality of the term `fitness' has led to misunderstandings and disputes. Even though we often may not see it happening, because of its slowness in terms of the timescale of our lives, natural selection never stops. Even humans are still evolving. For instance, our diet differs from that of our ancestors, and our teeth can function quite well on soft modern foods even if they are not very strong. The high sugar content of many modern foods leads to tooth decay, and potentially to abscesses that can be fatal, but there is no longer very pronounced natural selection for strong teeth, because dental care can solve these problems, or provide false teeth.

Just as for other functions that are no longer used intensively, changes are to be expected, and our teeth could one day become vestigial. They are already smaller than those of our close relatives, the chimpanzees, and there is no reason why they should not become smaller still. Excess sugar in the diet has also led to an increasing frequency of late-onset diabetes in human populations, with high mortality for sufferers. In the past, this disease was largely confined to people past childbearing age, but the age of onset is becoming steadily earlier. There is therefore a new, probably intense, selection pressure to change our metabolism so as to tolerate our changed diet. In an earlier chapter, we will show how changes in human life are leading to the evolution of greater longevity.

The concept of fitness is often misunderstood. When biologists try to illustrate the meanings of this term, they often use examples that correspond with our everyday use of the word fitness, such as the speed of antelopes. There is less danger of confusion if we think of characteristics like the lightweight bones of birds, with their hollow centers and strengthening cross-struts. The theory of natural selection accounts for such apparently well-designed structures by pointing out that, when flight was evolving, lighter-boned individuals would have had slightly higher chances of survival than others. If their descendants inherited lighter bones, the characteristic would increase in its representation in the population over the generations.

> *This is just the same as artificial selection by breeders of the fastest dogs, which has given all greyhounds long, thin legs. These are mechanically more efficient than short ones, and greyhounds' legs closely resemble those of antelopes and other fast-running animals, which have evolved by natural selection*

We can describe natural and artificial selection perfectly well without using the word fitness. Natural selection implies nothing more than those certain heritable variants may be preferentially passed on to future generations. Individuals carrying genes that lower their survival or reproductive success will generally not pass on those genes to the same extent as other individuals whose genes give higher survival or reproductive ability.

The term fitness is merely a useful short-cut to help briefly express the idea that characteristics sometimes affect organisms' chances of surviving and/or reproducing, without having to specify a particular characteristic. It is also useful in making mathematical models of the way selection affects the genetic make-up of a population. Conclusions from these models provide a rigorous underpinning for many of the statements that we make in this chapter, but we will not describe them here.

To illustrate selection of an advantageous mutation, consider the arms race between humans and rats, in which we try to develop rat poisons, and rats evolve resistance. The rat poison warfarin kills rats because it prevents blood clotting. It binds to an enzyme needed in the metabolism of vitamin K, which is important for blood clotting and many other functions. Resistant rats were once rare, because their vitamin K metabolism is changed, reducing growth and survival. In other words, there is a cost of resistance. In farms and towns where warfarin is used, however, only resistant animals can survive, so there is strong natural selection, despite the cost. The resistant version of the gene has therefore spread to high frequencies in the rat population, though the cost keeps it from spreading to all members of the species. However, a recent development is the evolution of a new kind of resistance which seems to be free from the cost, and which may even be advantageous (in the absence of poison). There is thus continued evolution in response to a change in the rats' environment. Variability and selection are very general properties of many systems, not just individual organisms. Certain components of the genetic material are maintained, not because they increase the fitness of the organisms that carry them, but because they can multiply within the genetic material itself, just like parasites in the body of their host. 50% of human DNA is thought to belong in this category. Another important situation in which natural selection drives evolutionary change within an organism occurs in cancers. Cancer is a disease in which a cell and its descendants evolve selfish behavior and multiply, regardless of the good of the rest of the body.

The disease is often caused by a mutation that increases the mutation rates of other genes (for instance, by a failure in the proofreading system described in Chapter 3, which checks DNA sequences and prevents mutations). If mutations occur at a high frequency, some may affect cell multiplication rates, and a fast-multiplying lineage may appear. As time goes on, more and more of the cells will descend from cells carrying mutations in other genes which confer faster and faster growth, and so the cancer often becomes more aggressive.

Cancer cells can also become resistant to drugs that suppress their growth. Like the well-known situation of drug-resistant HIV viruses evolving in an AIDS patient, cancer cells which acquire mutations that allow them to escape drug suppression outgrow the initial type of cells, and cause loss of remission of cancers. This is why it is often hopeless to restart drug treatment after a remission stops.

At the other extreme, there may be different rates of extinction of species with different sets of characteristics, that is there can be selection at the level of species. For example, species with large body sizes, which tend to have low population sizes and low rates of reproduction, are more vulnerable to extinction than species with small bodies.

In contrast, selection between individuals of the same species often favors larger body size, probably because larger individuals have greater success in competition for food or mates. The range of body sizes that we see in a group of related species may reflect the net outcome of both types of selection. Selection on individuals within species is likely, however, to be the most important factor, since it produces the different range of body sizes in the first place, and it usually operates much faster than selection at the species level. Selection is also important in non-biological contexts. In designing machines and computer programs, it has been found that a very efficient way to find the optimal design is to successively make small, random changes to the design, keeping versions that do the job well, and discarding others. This is increasingly being used to solve difficult design problems for complex systems. In this process, the engineer does not have a design in mind, but only the desired function.

9.16 Adaptations and Evolutionary History

The theory of evolution by natural selection explains features of organisms as a result of the successive accumulation of changes, each giving higher survival or reproductive success. What changes are possible depends on the pre-existing state of the organism: mutations can only modify the development of an animal or plant within certain limits, which are constrained by the underlying existing developmental programs that lead to the adult organism.

The results of artificial selection as practiced by animal and plant breeders show that it is relatively easy to change the sizes and shapes of body parts, or to produce striking changes in superficial characters such as external coloration, as in different breeds of dog. Radical changes can easily be produced by mutations, and laboratory geneticists have no difficulty in creating strains of mice or fruit flies that differ much more from normal forms than wild species differ from each other. It is possible, for example, to produce flies with four wings instead of the normal two. These major changes, however, often severely disrupt normal development, reducing survival and fertility, and are therefore unlikely to be favored by natural selection. They even tend to be avoided by animal and plant breeders (although such mutations have been used in developing unusual pigeon and dog breeds, where the animals' health is of lesser importance than for farmers).

For this reason, we expect that evolution will usually proceed by fairly small adjustments to what has gone before, rather than by sudden jumps to radically new states. This is particularly obvious for complex traits that depend on the mutual adjustment of many different components, such as the eye. If one component is changed drastically, it may not function well in combination with other parts that remain unchanged. When new adaptations evolve, they will usually be modified versions of pre-existing structures, and will at first often not be the optimum functional engineering design solutions. Natural selection resembles an engineer improving machinery by tinkering with it and modifying it, rather than sitting down and planning entirely new designs.

Modern screwdrivers can be suitable for precision work, with a diversity of heads suited for different purposes, but the evolutionary ancestors of screws were coarse-threaded spigots turned by a spike through a hole in one end. While we are often astonished by the precision and efficiency of adaptations of living organisms, there are many examples of tinkering, betrayed by features that make sense only in terms of their historical origins.

> *Painters represent angels with wings on their shoulders, allowing them the continued use of their arms. But the wings of all real flying or gliding species of vertebrates are modified forelimbs, so that pterodactyls, birds, and bats have all lost the use of their forelimbs for most of their original functions.*

9.17 Detecting Natural Selection

Darwin and Wallace argued that natural selection is the cause of adaptive evolution without knowing examples of selection operating in nature. Over the last 50 years, many cases of natural selection have been detected in action and studied in detail, immeasurably strengthening the evidence for its key role in evolution. We have space for only a few examples. A very important kind of natural selection acting today is causing ever-increasing antibiotic resistance in bacteria. This is an example of evolutionary change that is intensively studied, because it endangers our lives, and occurs fast and (unfortunately) very repeatedly. On the day we were writing this, the headlines in the newspaper were about methicillin resistant Staphylococcus in Edinburgh's Royal Infirmary. Whenever an antibiotic is widely used, resistant bacteria are soon found. Antibiotics were first widely used in the 1940s, and concerns about resistance were soon being raised by microbiologists. In 1955, an article in the American Journal of Medicine, aimed at doctors, wrote that the indiscriminate use of antibiotics: 'is fraught with the risk of selecting resistant strains', and in 1966 (when people had not changed their behavior), another microbiologist wrote: 'is there no way to generate sufficient general concern so that antibiotic resistance can be attacked?'

The speedy evolution of antibiotic resistance is not surprising, because bacteria multiply fast and are present in enormous numbers, so that any mutation that can make a cell resistant is sure to occur in a few bacteria in a population; if the bacteria are able to survive the change to their cell functions caused by the mutation and to multiply, a resistant population can rapidly build up. One might hope that resistance will be costly for bacteria, as was initially true for warfarin resistance in rats, but as in rats we cannot rely on this remaining true for long. Sooner or later, bacteria will evolve so that they survive well in the presence of antibiotics, without serious costs to themselves. Our only chance is therefore to use antibiotics sparingly, confining use to situations where they are really needed, and making sure that all infecting bacteria are killed quickly, before they have time to evolve resistance.

If one stops treatment while some bacteria remain present, their population will inevitably include some resistant bacteria, which can then spread to other people. Antibiotic resistance can also spread between bacteria, even ones of different species. Antibiotics given to farm animals, to keep infections down and promote growth, can cause resistance to spread to human pathogens. Even these consequences are not the whole of the problem. Bacteria that have resistance mutations are not typical of their populations, but sometimes have higher mutation rates than average, allowing them to respond even faster to selection. Drug and pesticide resistance evolve whenever drugs are used to kill parasites or pests, and literally hundreds of cases have been studied in microbes, plants, and animals. Even the HIV virus mutates within AIDS patients treated with drugs and evolves resistance so that the treatment eventually fails. To try to prevent this, two drugs instead of one are often used. Because mutations are rare events, the virus population in a patient is unlikely to get both resistance mutations very quickly, but this usually happens eventually.

A recent study showed that the sequence divergence for 53 noncoding DNA sequences compared between humans and chimpanzees ranged between 0 and 2.6% of the total letters and averaged only 1.24% (1.62% for the human and gorilla). These estimates show why it is now accepted that chimpanzees, rather than gorillas, are our closest living relatives.

The differences are much greater if humans are compared with the orangutan, and greater still if we are compared with baboons. More distantly related mammals, such as carnivores and rodents, differ at the sequence level much more than do different primates; mammals differ much more from birds than they do from each other, and so on. The patterns of relationships revealed by sequence comparisons are in broad agreement with what is expected from the times at which the major groups of animals and plants appear in the fossil record, as expected in the theory of evolution.

The table of sequence differences shows that silent changes are generally much more common than replacement changes, although even silent changes are rare between the most closely related species such as chimpanzees and humans. The obvious interpretation is that most changes to the amino acid sequence of a protein impair its function to some extent.

As we described in an earlier chapter, a small detrimental effect caused by a mutation will result in selection quickly eliminating the mutation from the population. Most mutations that change protein sequences therefore never contribute to evolutionary differences in gene sequences that accumulate between species. But there is also increasingly firm evidence that some amino acid sequence evolution is driven by selection acting on occasional favorable mutations, so that molecular adaptation occurs.

In contrast to the often-detrimental effects of mutations changing amino acids, silent changes to the sequences of genes will have little or no effect on biological functions. It thus makes sense that most divergence in gene sequences between species are silent changes. But when a new silent mutation appears in a population, it is just a single copy among thousands or millions of copies of the gene in question (two in each individual in the population). How does such a mutation spread through the population if it does not confer any selective advantage to its carrier? The answer is that random changes in the frequencies of alternative variants (genetic drift) take place in finite populations, a concept we introduced briefly in an earlier chapter.

This process works as follows. Suppose that we study a population of the fruit fly Drosophila melanogaster. For the population to be maintained, each adult must contribute on average two descendants to the next generation.

Suppose that the population varies in eye color, with some individuals carrying a mutant gene that makes the eyes bright red while the non-mutant version of this gene makes all the other flies' eyes the normal dull red. If individuals with either type of gene have the same average number of offspring, there is no selection on eye color; it is said to be neutral in its effects. Because of this neutrality with respect to selection, the genes of the next generation will be drawn randomly from the parental population. Some individuals may have no offspring, while others may happen by chance to have more than the average of two offspring. This means that the frequency of the mutant gene in the progeny generation will not be the same as its frequency among the parents, since it is extremely unlikely that individuals with and without the mutant gene contribute exactly the same numbers of offspring.

Over the generations, there will thus be continual random fluctuations in the composition of the population, until sooner or later either all members of the population have the gene for bright red eyes, or else it is lost from the population, and they all have the alternative version of the gene. In a small population, genetic drift is fast, and it will not take long until all members of the population become the same. This will take much longer in a large population.

This illustrates two effects of genetic drift. First, while a new variant is drifting to eventual loss or to a frequency of 100% (fixation), the character affected by the gene is variable within the population. The input of new neutral variants by mutation and the changes in variant frequencies (and, from time to time, loss of variant genes) by drift determines the variability in the population. Examination of DNA sequences of the same gene from different individuals from a population reveals variability at silent sites due to this process, as we mentioned in an earlier chapter.

A second effect of genetic drift is that a selectively neutral variant that is initially very rare has some chance of spreading throughout the whole population and replacing alternative variants, although it has a much greater chance of being lost. Genetic drift thus leads to evolutionary divergence between isolated populations, even without any selection promoting the changes.

This is a very slow process. Its rate depends on the rate at which new neutral mutations arise, as well as the rate at which genetic drift leads to replacement of one version of a gene by a new one. Remarkably, it turns out that the rate of DNA sequence divergence between a pair of species depends only on the rate of mutation per DNA letter (the frequency with which a particular letter in a parent is mutant in the copy that is passed to an offspring). An intuitive explanation for this is that, if no selection is acting, nothing affects the number of mutational differences between a pair of species except the rate at which mutations appear in the sequence and the amount of time since the species' last common ancestor. A large population has more new mutations per generation, simply because there are more individuals in which a mutation might happen.

But genetic drift happens faster in a small population, as explained above. It turns out that the two opposing effects of population size cancel out exactly, and so the mutation rate determines the rate of divergence.

9.18 Genetic Drift

The process of genetic drift is the drift or change of a single gene over six generations, in a population of five individuals. Each individual (symbolized by an open shape) has two copies of the gene, one from each parent. The different DNA sequences of the individuals' gene copies are not shown in detail but are symbolized by black discs with or without a white spot. The white spots might correspond to the variant gene causing bright red eye color, and the black discs to the variant with dull red eye color, in the Drosophila example given in the text. In the first generation, three individuals have one of the white spot types of the gene and one of the black type. Thus 30% of the genes in the population have the white spot. The figure shows the lines of descent of the genes in each generation (we assume for convenience that individuals can reproduce as either male or female, as is true for many hermaphroditic species of plants, such as tomatoes, and some animals, such as earthworms). Some individuals happen by chance to have more offspring than others, while other have less, or may even leave no surviving descendants (e.g. the individual shown at the right in generation 2).

The numbers of white spot and black gene copies therefore fluctuate from each generation to the next. In the third generation, three individuals all inherit the white spot gene copy from the single individual carrying one such gene in generation 2, so this type of gene goes from 10% to 30%; in the next generation it is 50%, and so on.

This theoretical result has important implications for our ability to determine the relationships between different species. It implies that neutral changes accumulate in a gene as time goes on, at a rate that depends on the gene's mutation rate (the molecular clock principle, which we mentioned, but did not explain this change). Sequence changes in genes are therefore likely to take place in a much more clock-like fashion than changes in traits subject to selection.

Rates of morphological changes depend strongly on environmental changes, and variable rates and reversals of direction can occur. Even the molecular clock is not very precise. Rates of molecular evolution can change over time within the same lineage, as well as between different lineages. Nevertheless, use of the molecular clock allows biologists to roughly date the divergence between species for which there is no fossil evidence. To calibrate the clock, one needs sequences from the closest available species whose divergence dates are known.

One of the most important applications of this method has been to date the timing of the split between the lineage giving rise to modern humans and the one leading to chimpanzees and gorillas, for which no independent fossil evidence is available. Use of the molecular clock with a large number of gene sequences has enabled a date of 6 or 7 million years to be estimated with considerable confidence. Because the rate of neutral sequence evolution depends on the mutation rate, the clock is exceedingly slow since the rate at which single letters in the DNA change by mutation is very low.

The fact that approximately 1% of the DNA letters differ between humans and chimpanzees corresponds to a single letter changing only once in over a billion years. This is consistent with experimental measurements of mutation rates.

A molecular clock is also found to apply to the amino acid sequence of proteins. As already mentioned, protein sequences evolve more slowly than silent DNA differences, and are therefore useful for the difficult task of comparing species that diverged a very long time ago. Between such species, multiple changes will have occurred at some sites in their DNA sequences, so that it becomes impossible to count accurately the number of mutations that have happened. Scientists who are interested in reconstructing the times of divergence between the major groups of living forms therefore use data from slowly evolving molecules. Such dates are, of course, rough estimates, but the accumulation of estimates from many different genes can improve the accuracy of the procedure.

Judicious use of sequence information from genes that evolve at different rates is allowing evolutionary biologists to form a picture of the relationships between groups of organisms whose last common ancestors lived a billion or more years ago. In other words, we are getting close to reconstructing the genealogical tree of life.

9.19 How Can Complex Adaptations Evolve?

Critics of the theory of evolution by natural selection frequently raise the difficulty of evolving complex biological structures, from protein molecules through single cells to eyes and brains. How can a fully functioning and beautifully adapted piece of biological machinery be produced purely by selection acting on chance mutations? The key to understanding how this can happen is expressed in another meaning of the word `adapt'. In the evolution of organisms and their complex machinery, many aspects are modified (adapted) versions of pre-existing structures, just as when machines are made by engineers. In making complex machines and devices, less elegant initial models are refined over the course of time and diversified (adapted) to new, sometimes unanticipated, uses. The evolution of the total knee replacement is a good example of the process by which a crude initial solution to a problem was good enough to be useful but was successively adapted to work better and better. Just as in biological evolution, many early designs were developed that seem poor by today's standards, yet each was an improvement on the ones before, and could be used by knee surgeons. These each played their roles as stages in the evolution of modern, complex artificial knees. This process of successive adaptation of `designs' is like climbing a hill in a thick fog. Even without a goal of reaching the top (or even without knowing where it is), if one follows a simple rule - each step goes uphill - one will move closer and closer to the summit (or at least to a local top). Simply by making a structure work better in one way or another, the end result is an improved design, without a Designer being necessary. In engineering, improved design is often the result of many contributions from different engineers over the evolution of a machine, and early car designers would have been astonished at modern cars.

In natural evolution, it results from what has been called 'tinkering' with the organism, with minor changes that make their possessors survive or reproduce better than others.

In the evolution of a complex structure, several different traits must, of course, evolve simultaneously, so that the different parts of the structure are well adapted to function as a whole. We saw in an earlier chapter that advantageous traits can spread through a population over a short time, relative to the time available for major evolutionary changes, even if they are initially very rare. A succession of small changes to a structure that already works, but can be improved, can therefore produce large evolutionary changes. After many thousands of years, the radical transformation of even a complex structure is not difficult to imagine. After enough time, the structure will differ from its ancestral state in many different ways, so that individuals in the descendant population would have combinations of characteristics never seen in the ancestral population, just as modern cars have many differences from early cars. This is not just a theoretical possibility: as we described in an earlier chapter. Animal and plant breeders routinely accomplish this by artificial selection. There is thus no difficulty in seeing how natural selection can cause the evolution of highly complex characters, made up of numerous mutually adjusted components.

The evolution of protein molecules is sometimes posed as an especially difficult problem. Proteins are complex structures whose parts must interact to function properly (many proteins must also interact with other proteins and other molecules, including DNA in some cases). The theory of evolution must certainly be capable of accounting for protein evolution.

There are 20 different kinds of amino acids, so the chance that the right one would appear at a particular site in a protein molecule 100 amino acids long (shorter than many real proteins) is 1 in 20. The chance is evidently vanishingly small that, if 100 amino acids were randomly thrown together, each position in the sequence would have the right amino acid, and a working protein would form. It has therefore been claimed that the chance of assembling a functioning protein is similar to that of an airliner being assembled by a tornado blowing through a scrap yard.

It is true that a functioning protein could not be assembled by randomly picking an amino acid for each position in the sequence. But, as the explanation given above makes clear, natural selection does not work like this. Proteins probably started as short chains of a few amino acids that could cause reactions to go a bit faster and were successively improved as they evolved. There is no need to worry about the many millions of potential non-functional sequences that will never exist, provided that protein sequences during evolution started off catalysing reactions better than when no protein is present, and then got successively better over evolutionary time. It is easy to see in principle how successive stepwise changes, each one changing the sequence or adding to its length, could improve a protein.

Our knowledge about how proteins function supports this. The part of a protein that is essential for its chemical activity is often only a very small part of its sequence. A typical enzyme has just a handful of amino acids that physically interact with the chemical that is to be changed by the enzyme. Most of the rest of the protein chain simply provides a scaffold that supports the structure of the part involved in this interaction. This implies that the functioning of a protein depends critically on only a relatively small set of amino acids, so that a new function could evolve by a small number of changes to the sequence of the protein. This has been verified by numerous experiments in which artificially induced changes to protein sequences have been subjected to selection for new activities. It has proved surprisingly easy to produce quite radical shifts in the biological activity of proteins by these means, sometimes just by a change in a single amino acid, and there are similar examples among naturally evolved changes.

If complex adaptations really evolve in steps, as evolutionary biologists propose, we should be able find evidence for intermediate stages in the evolution of such characters. There are two sources of such evidence: the existence of intermediates in the fossil record, and present-day species that show intermediate stages between simple and more advanced states. In an earlier chapter, we described examples of intermediate fossils linking very different forms; these support the principle of stepwise evolutionary changes. Of course, in many cases there is a complete absence of intermediates, especially as we go further back in time. In particular, the major divisions of multi-cellular animals, including molluscs, arthropods, and vertebrates, nearly all appeared rather suddenly in the Cambrian (more than 500 million years ago), with virtually no fossil evidence concerning their ancestors. Recent DNA sequence studies of the relationships between them suggests strongly that these groups were already separate lineages long before the Cambrian era, But we simply have no information on what they looked like, probably because they were soft-bodied and hence unlikely to fossilize. But the incompleteness of the fossil record does not mean that intermediates did not exist. New intermediates are constantly being discovered. A recent one is a 125-million-year-old mammalian fossil from China with features similar to those of modern placental mammals, but more than 40 million years older than the oldest previously known fossil of this kind.

The other type of evidence, from comparisons of living forms, is our only source of information on features that do not fossilize. A simple but compelling example is provided by flight, as pointed out by Darwin in an earlier chapter of The Origin of Species. There are no fossils connecting bats with other mammals; the first bat fossils, found in deposits over 60 million years old, have the same highly modified limbs as modern bats. But there are several examples of modern mammals which have the ability to glide but cannot fly. The most familiar are the `flying' squirrels, which are very similar to ordinary squirrels except for flaps of skin connecting their fore- and hindlimbs. These act as a crude wing, which allows the squirrels to glide some distance if they launch themselves into the air. Similar adaptations to gliding have evolved independently in other animals.

CHAPTER 10
HUMAN EVOLUTION – ARCHAIC & MODERN HOMINIDS
10.1 A Root of the Tree of Life Leading to our Ancestors

Many of the important advances made by biologists in the past 150 years can be reduced to a single metaphor. All living, or extant, organisms, that is, animals, plants, fungi, bacteria, viruses, and all the types of organisms that lived in the past, are situated somewhere on the branches and twigs of an arborvitae or Tree of Life. We are connected to all organisms that are alive today, and all the organisms that have ever lived, via the branches of the Tree of Life (TOL). The extinct organisms that lie on the branches that connect us to the root of the tree are our ancestors. The rest, on branches that connect directly with our own, are closely related to modern humans, but they are not our ancestors. The long' version of human evolution would be a journey that starts approximately three billion years ago at the base of the TOL with the simplest form of life. We would then pass up the base of the trunk and into the relatively small part of the tree that contains all animals, and on into the branch that contains all the animals with backbones. Around 400 million years ago we would enter the branch that contains vertebrates that have four limbs, then around 250 million years ago into the branch that contains the mammals, and then into a thin branch that contains one of the subgroups of mammals called the primates. At the base of this primate branch we are still at least 50–60 million years away from the present day. The next part of this long' version of the human evolutionary journey takes us successively into the monkey and ape, the ape and then into the great ape branches of the Tree of Life. Sometime between 15 and 12 million years ago we move into the small branch that gave rise to contemporary modern humans and to the living African apes. Between 11 and 9 million years ago the branch for the gorillas split off to leave just a single slender branch consisting of the ancestors of both extant (i.e. living) chimpanzees and modern humans. Around 8 to 5 million years ago this very small branch split into two twigs. One of the twigs ends on the surface of the TOL with the living chimpanzees, the other leads to modern humans. Paleoanthropology is the science that tries to reconstruct the evolutionary history of this small, exclusively human, twig.(Wood)

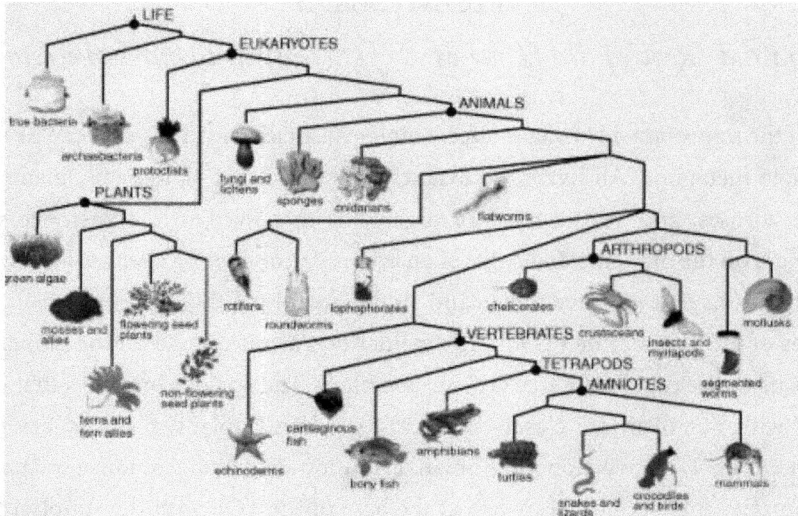

Figure 10-1 :Tree of life - Classification of organisms into the hierarchical system

The next part of this long' version of the human evolutionary journey takes us successively into the monkey and ape, the ape and then into the great ape branches of the Tree of Life. Sometime between 15 and 12 million years ago we move into the small branch that gave rise to contemporary modern humans and to the living African apes. Between 11 and 9 million years ago the branch for the gorillas split off to leave just a single slender branch consisting of the ancestors of both extant (i.e., living) chimpanzees and modern humans. Around 8 to 5 million years ago this very small branch split into two twigs. One of the twigs ends on the surface of the TOL with the living chimpanzees, the other leads to modern humans. Paleoanthropology is the science that tries to reconstruct the evolutionary history of this small, exclusively human, twig.

This chapter focuses on the last stage of the human evolutionary journey, the part between the most recent common ancestor shared by chimpanzees and humans and present-day modern humans. To understand this, we need to use some scientific jargon. So instead of referring to twigs' we need to use the proper biological term clade': extinct side branches are called subclades'.

Species anywhere on the main human twig, or on its side branches, are called hominins'; the equivalent species on the chimp twig are called panins'. And instead of writing out millions of years' and millions of years ago' (and the equivalents for thousands of years) we will use instead the abbreviations MY and MYA' and KY' and KYA'. This Very Short Introduction has three objectives. The first is to try and explain how paleoanthropologists go about the task of improving our understanding of human evolutionary history. The second is to convey a sense of what we think we know about human evolutionary history, and the third is to try to give a sense of where the major gaps in our knowledge are.

We use two main strategies to improve our understanding of human evolutionary history. The first is to obtain more data. You can get more data by finding more fossils, or by extracting more information from the existing fossil evidence. You can find more fossils from existing sites, or you can look for new sites. You can extract more information from the existing fossil record by using techniques such as confocal microscopy and laser scanning to make more precise observations about their external morphology. You can also gather information about the internal morphology and biochemistry of fossils. This ranges from using non-invasive medical imaging techniques such as computed tomography to obtain information about structures like the inner ear, to using new types of microscopes to investigate the microscopic anatomy of teeth, and the latest molecular biology technology to detect small amounts of DNA in fossils. The second strategy for reducing our ignorance about human evolutionary history is to improve the ways we analyze the data we do have. These improvements range from more effective statistical methods to the use of novel methods of functional analysis.

Researchers also try to improve the ways they generate and test hypotheses about the numbers of species in the hominin fossil record, and about how those species are related to each other and to modern humans and chimpanzees.

2. Progress has can be made in paleoanthropology research such as collecting and extracting more information from existing fossil and improving analytical methods. Historically philosophers and then scientists came to realize that modern humans are part of the natural world. Today scientists think chimpanzees are more closely related to modern humans than they are to gorillas, and the chimp/human common ancestor lived between 8 and 5 MYA. In this chapter, the lines of evidence are used to investigate what the 8–5 MY-old hominid clade looks like. Is it bushy', or straight like the stem of a thin spindly plant?

How much of it can be reconstructed by looking at variation in modern humans, and what needs to be investigated by searching for, finding, and then interpreting fossil and archaeological evidence? Where do researchers look for new fossil sites, and how do they date the fossils they find? Researchers decide how many species there are within the hominin clade. I also review the methods researchers use to determine how many hominin subclades there are, and how they are related to one another. After that we consider possible' and probable' early hominins. The chapter reviews four collections of fossils that represent each of the candidate' taxa that have been put forward for being at the very base of the hominin clade. Then later in the chapter we will look at archaic' and transitional' hominins. These are fossil taxa that almost certainly belong to the hominin clade, but which are still a long way from being like modern humans. The next chapter looks at hominins researchers believe might be the earliest members of the genus Homo: we call these pre-modern' Homo. I look at the earliest fossil evidence of pre-modern Homo from Africa, and then follow Homo as it moves out of Africa into the rest of the Old World. The next chapter considers evidence about the origin and subsequent migrations of anatomically modern humans, or Homo sapiens. When and where do we find the earliest fossil evidence of anatomically modern humans?

Did the change from pre-modern Homo to anatomically modern humans happen several times and in several different regions of the world? Or did anatomically modern humans emerge just once, in one place, and then spread out, either by migration or by interbreeding, so that modern humans eventually replaced regional populations of pre-modern Homo? Finally, what will not be in this book? This chapter will concentrate on the physical and not the cultural aspects of human evolution.

10.2 Finding Our Place

Long before researchers began to accumulate material evidence about the many ways modern humans resemble other animals, and long before Charles Darwin and Gregor Mendel laid the foundations of our understanding of the principles and mechanisms that underlie the connectedness of the living world, Greek scholars had reasoned that modern humanity was part of, and not apart from, the natural world.

When did the process of using reason to try and understand human origins begin, and how did it develop? When was the scientific method first applied to the study of human evolution?

Plato and Aristotle in the 5th and 6th BCE provide the earliest recorded ideas about the origin of humanity. These early Greek philosophers suggested that the entire natural world, including modern humans, forms one system. This means that modern humans must have originated in the same way as other animals. The Roman philosopher Lucretius, writing in the 1st century BCE, proposed that the earliest humans were unlike contemporary Romans. He suggested that human ancestors were animal-like cave dwellers, with neither tools nor language. Both classical Greek and Roman thinkers viewed tool and fire making and the use of verbal language as crucial components of humanity. Thus, the notion that modern humans had evolved from an earlier, primitive form was established early on in Western thought.(Wood)

10.3 Reason is Replaced by Faith

After the collapse of the Roman Empire in the 5th century Graeco Roman ideas about the creation of the world and of humanity were replaced with the narrative set out in Genesis: reason-based explanations were replaced by faith-based ones. The main parts of the narrative are well known. God created humans in the form of a man, Adam, and then a woman, Eve. Because they were the result of God's handiwork Adam and Eve must have come equipped with language and with rational and cultured minds.

According to this version of human origins, the first humans were able to live together in harmony, and they possessed all the mental and moral capacities that, according to the biblical narrative, set humanity above and apart from other animals. The biblical explanation for the different races of modern humans is that they originated when Noah's offspring migrated to different parts of the world after the last big biblical flood, or deluge. The Latin for flood' is diluvium, so we call anything very old antediluvial', or dating from before the flood'. Explanations for the creation of the living world involving successive floods had implications for the science that was to become known as paleontology.

All the animals created after a flood must inevitably perish at the time of the next flood. Thus antediluvial' animals should never coexist with the animals that replaced them. We will return to this and other implications of diluvialism later in this chapter. The Bible also has an explanation for the rich variety of human languages. It suggests that God wanted to promote confusion among the people constructing the tower of Babel, and that he did so by creating mutually incomprehensible languages.

In the Genesis version of human origins, The Devil's successful temptation of Adam and Eve in the Garden of Eden forced them and their descendants to learn afresh about agriculture and animal husbandry. They had to reinvent all the tools needed for civilized life. With very few exceptions Western philosophers living in and immediately after the Dark Ages (5th to 12th centuries) supported a biblical explanation for human origins. This changed with the rediscovery and rapid growth of natural philosophy that was only later called science. But, paradoxically, not long after the scientific method began to be applied to the study of human origins in the 19th and 20th centuries some religious groups responded to attempts by scientists to interpret the Bible less literally by being even stricter about their biblical literalism. This reaction was the origin of creationism, and of what, erroneously, is called Creation Science'. During the Dark Ages very few Greek classical texts survived in Europe. The few that did survive were read and valued by Muslim philosophers and scholars, and some of them were translated into Arabic.

When the Muslims were driven out of Spain in the 12th century, a few medieval Christian scholars were curious enough to translate these manuscripts from Arabic into Latin. Some of these translated texts dealt with the natural world, including human origins.

For example, the 13th-century Italian Christian philosopher, Thomas Aquinas, integrated Greek ideas about nature and modern humans with some of the Christian interpretations based on the Bible.

The work of Thomas Aquinas and his contemporaries laid the foundations of the Renaissance when science and rational learning were reintroduced into Europe. Science re-emerges The move away from reliance on biblical dogma was especially important for those who were interested in what we now call the natural sciences, such as biology and the earth sciences.

An Englishman, Francis Bacon, was a major influence on the way scientific investigations developed. Theologians use the deductive method: beginning with a belief, they then deduce the consequences of that belief. Bacon suggested that scientists should work in a different way he called the inductive' method. Induction begins with observations, also called evidence or data'. Scientists devise an explanation, called a hypothesis', to explain those observations. Then they test the hypothesis by making more observations, or in sciences like chemistry, physics and biology, by conducting experiments. This inductive way of doing things is the way the sciences involved in human evolution research are meant to work. Bacon summarized his suggestions about how the world should be investigated in aphorisms and set these out in his book called the Novum Organum or true suggestions for the interpretation of Nature, published in 1620. His message was a simple one. Do not be content with reading about an explanation in a book. Go out, make observations, investigate the phenomenon for yourself, then devise and test your own hypotheses.

Anatomy starts to become scientific Nearly three-quarters of a century before Bacon published this advice, a major change had already occurred in anatomy, the natural science closest to the study of human evolution. That change was the work of Andreas Vesalius. Born in 1514 in what is now Belgium, Vesalius finished his medical studies in 1537. In the same year he was appointed to teach anatomy and surgery in Padua, Italy. Vesalius' own anatomy education was typical for the time.

The professor sat in his chair (hence professorships are called chairs') and read out loud from the only locally available textbook. He sat at a safe distance from a human body that was being dissected by his assistant.

It did not take long for Vesalius to realize that he and his fellow students were being told one thing by their professor and were being shown something else by the professor's assistant.

In 1540 Vesalius visited Bologna where, for the first time, he was able to compare the skeletons of a monkey and a human. He realized the textbooks used by his professors were based on a confusing mixture of human, monkey, and dog anatomy, so he resolved to write his own, accurate, human anatomy book. The result, the seven-volume De Humani Corporis Fabrica Libri Septem, or On the Fabric of the Human Body, was published in 1543. Vesalius performed the dissections and sketched the drafts of the illustrations: the Fabrica is one of the great achievements in the history of biology. Vesalius' successful efforts to make anatomy more rigorous ensured that scientists would have access to reliable information about the structure of the human body.

Geology emerges Another field of science relevant to the eventual study of human origins, geology (now usually referred to as earth science'), developed more gradually than anatomical science. One of the implications of interpreting the Genesis narrative literally is that the world, and therefore humanity, cannot have had a long history. There is a long tradition of biblically based chronologies, beginning with people like Isidore of Seville and the Venerable Bede in the 6th and 7th centuries, respectively. The one cited most often was published in 1650 by James Ussher, then archbishop of Armagh in Ireland. He used the number of begats' in the Book of Genesis to calculate the precise year of the act of Creation, which, according to his arithmetic, was in 4004 BC.

Subsequently, another theologian John Lightfoot, of Cambridge University, England, refined Ussher's estimate and declared that the act of Creation took place precisely at 9 a.m. on 23 October 4004 BCE. Geology, and especially the work of James Hutton, provided an alternative calendar, suggesting the earth and its inhabitants were substantially older than this. The development of geology was substantially influenced by the Industrial Revolution.

The excavations involved in making cuttings' for canals and railroads gave amateur geologists the opportunity to see previously hidden rock formations. Pioneer geologists such as William Smith and James Hutton paved the way for Charles Lyell in 1830 to set out a rational version of the history of the earth in The Principles of Geology. Lyell's book influenced many scientists, including Charles Darwin, and it helped establish fluvialism and uniformitarianism as alternatives to biblically based diluvial explanations for the state of the landscape.

Fluvialism suggested that erosion by rivers and streams had reduced the height of mountains and created valleys and thus played a major role in shaping the contours of the earth. Uniformitarianism suggested that the processes that shaped the earth's surface in the past, such as erosion and volcanism, were the same processes we see in action today. Lyell also championed the principle that rocks and strata generally increase in age the further down they are in any relatively simple geological sequence. Barring major and obvious upheavals and deliberate burial, the same principle must apply to any fossils or stone tools contained within those rocks. The lower in a sequence of rocks a fossil is, the older it is likely to be. The implications of the new science of geology were profound. There was no need to invoke the biblical floods or divine intervention to explain the appearance of the earth. The pioneer geologists of the time also suggested that it would have taken the processes that are shaping the earth's surface today a lot longer than the 6,000 years implied by the Genesis narrative to make the changes the pioneer geologists had observed. (Wood)

10.4 Fossils

Classical Greek and Roman writers had recognized the existence of fossils, but they mostly interpreted them as remnants of the ancient monsters that figure prominently in their myths and legends. By the 18th century, geologists began to accept that life-like structures in rocks were the remains of extinct animals and plant and that there was no need to invoke supernatural reasons for their existence. The association of the fossil evidence of exotic extinct animals with creatures closely related to living forms in the same strata effectively refuted the diluvial theory, for as I noted earlier in the chapter the latter does not allow for any mixing of modern and ancient, or antediluvial, animals. In addition to the important conclusions reached by pioneer geologists about the history of the earth, several other factors influenced 17th- and 18th-century scientists to consider alternatives to the Genesis account of human origins.

Explorers were returning from distant lands with eye-witness accounts of modern humans living in crude shelters, using simple tools, and existing by hunting and gathering. This was so far from the state of humanity in their homeland that European travelers described the people they observed as living in a state of savagery'.

According to the Genesis narrative, no human beings created by God should be living in such a state.

10.5 A Catalogue of life

The same explorers and traders who had returned to Europe with tales of the behavior of primitive people also brought back descriptions and sometimes suitably preserved specimens of many exotic plants and animals. When these discoveries were added to the more familiar plants and animals from Europe, they made for a perplexing array of plant and animal life. The living world badly needed a system for describing and organizing it. Several schemes were put forward, notably one by John Ray who introduced the concept of the species. However, the one that has stood the test of time was devised by a Swede called Karl von Linné, a name we know better in its Latinized form, Carolus Linnaeus.

Classification schemes try to group similar things together in increasingly broad, or inclusive, categories. Think of the following example of a classification of automobiles. It has seven levels or categories; it begins with the most inclusive category and ends with a small group. The levels are Vehicles', Powered Vehicles', Automobile', Luxury Car', Rolls-Royce', Silver Shadow', and 1970 Silver Shadow II'. The Linnaean classification system also recognizes seven basic levels. The most inclusive category, the equivalent of Vehicles' in our example, is the kingdom, followed by the phylum, class, order, family, genus, with the species being the smallest, least inclusive, formal category. Linnaeus' original seven-level system has been expanded by adding the category tribe' between the genus and family, and by introducing the prefix super-above a category, and the prefixes sub- and infra-, below it. These additions increase the potential number of categories below the level of order to a total of 12. The groups recognized at each level in the Linnaean hierarchy are called taxonomic groups'.

Each distinctive group is called a taxon' (pl. taxa'). Thus, the species Homo sapiens is a taxon, and so is the order Primates. When the system is applied to a group of related organisms, the scheme is called a Linnaean taxonomy, usually abbreviated to a taxonomy.

The Linnaean taxonomic system is also known as the binomial system because two categories, the genus and species, make up the unique Latinized name (e.g. Homo sapiens = modern humans; Pantroglodytes = chimpanzees) we give to each species. You can abbreviate the name of the genus, but not the species. So you can write H. sapiens and P. troglodytes, but not Homo s. or Pan t., as there can sometimes be more than one species name in that genus that begins with the same first letter, such as Homo sapiens and Homo soloensis.

10.6 Evidence of Connections

Trees are common metaphors. In religion, for example in Christianity, the Great Chain of Being is sometimes represented as a tree. Modern humans are on top of the tree, with other living animals placed within the tree at heights corresponding to their level of complexity. However, in contemporary life sciences the Tree of Life is not a metaphor: it is taken more literally. In a modern scientific Tree of Life, the relative size of the part of the tree given over to any particular group of living things reflects the number of taxa, and the pattern of branching within the tree reflects the way scientists think plants and animals are related. When the first science-based Trees of Life were constructed in the 19th century, the closeness of the relationship between any two animals had to be assessed using morphological evidence that could be studied with the naked eye or with a conventional light microscope.

The assumption was that the larger the number of shared structures the closer their branches will be within the TOL. Developments in biochemistry during the first half of the 20th century meant that, in addition to this traditional morphological evidence, scientists could use evidence about the physical characteristics of molecules. The earliest attempts to use biochemical information for determining relationships used protein molecules found on the surface of red blood cells and in plasma.

Both these lines of evidence emphasized the closeness of the relationship between modern humans and chimpanzees. Proteins are the basis of the machinery that makes other molecules, like sugars and fats, and ultimately the tissues that make up the components of our bodies, such as muscles, nerves, bones and teeth. In 1953 James Watson and Francis Crick, with the help of Rosalind Franklin, discovered that the nature of proteins, the building blocks of our bodies, is determined by the details of a molecule called DNA (short for deoxyribose nucleic acid). Scientists have shown since that DNA transmitted from parents to their offspring contains coded instructions, called the genetic code. This, in large measure, determines what the bodies of those offspring will look like.

These developments in molecular biology meant that instead of working out how species are related by comparing traditional morphology, or by looking at the morphology of protein molecules, scientists could determine relationships by comparing the DNA that dictates the structure and shape of proteins.

When these methods, first traditional anatomy, then the morphology of protein molecules, and finally the structure of DNA (the details of how DNA is compared are given below) were applied to more and more of the organisms in the Tree of Life it became apparent that animal species that were similar in their anatomy also had similar molecules and similar genetic instructions. Researchers have also shown that, even though the wing of an insect, and the arm of a primate look very different, the same basic instructions are used during their development. This is additional compelling evidence that all living things are connected within a single Tree of Life. The only explanation for this connectedness that has withstood scientific scrutiny is evolution; the only mechanism for evolution that has withstood scientific scrutiny is natural selection.

10.7 Evolution – An Explanation for the Tree of Life

Evolution means gradual change. In the case of animals this usually (but not always) means a change from a less complex animal to a more complex animal. We now know that most of these changes occur during speciation, which is when an old' species changes quite rapidly into a new', different, species.

Although the Greeks were comfortable with the idea that the behavior of an animal could change, they did not accept that the structure of animals, including humans, had been modified since they were spontaneously generated. Indeed, Plato championed the idea that living things were unchanging, or immutable, and his opinions influenced philosophers and scientists until the middle of the 19th century. A French scientist, Jean Baptiste Lamarck, in his Philosophie Zoologique published in 1809, set out the first scientific explanation for the Tree of Life. In the English-speaking world Lamarck's ideas were popularized in an influential book called Vestiges of the Natural History of Creation (1844).

We know that Vestiges influenced the two men, Charles Darwin, and Alfred Russel Wallace, who, independently, hit upon the concept that the main mechanism driving evolution was natural selection. Charles Darwin's contributions to science did not include the idea of evolution. What Darwin contributed was a coherent theory about the way evolution could work. As we will see, Darwin's theory of natural selection accounts for both the diversity and the branching pattern of the Tree of Life. Other books that influenced Darwin's thinking was Robert Malthus's Essay on the Principle of Population (1798) and Charles Lyell's Principles of Geology. Malthus stressed that resources are finite, and this suggested to Darwin that imbalances between the resources available and the demand for them might be the driving force behind the selection needed to make evolution happen. Lyell's fluvial explanation for the evolution of the surface of the earth was much like the gradual morphological change that Darwin suggested was responsible for the modification of existing species to produce new ones. Darwin was also goaded into action by the work and philosophy of William Paley. Paley was a champion of the notion that animals were so well adapted for their habitat that this cannot have been due to chance. He suggested that they must have been designed, and if so, there must be a designer, and that the designer must have been God.

Paley provoked Darwin to think about an alternative to the former's creationist interpretations. Charles Darwin made two seminal contributions to evolutionary science. The first was the recognition that no two individual animals are alike: they are not perfect copies. Darwin's other related contribution was the idea of natural selection.

In a nutshell, natural selection suggests that, because resources are finite, and because of random variation, some individuals will be better than others at accessing those resources. That variant will then gain enough of an advantage that it will produce more surviving offspring than other individuals belonging to the same species. Biologists refer to this advantage as an increase in an animal's fitness'. Darwin's notebooks are full of evidence about the effectiveness of the type of artificial selection used by animal and plant breeders. Darwin's genius was to think of a way that the same process could occur naturally.

Selection, and thus evolution, will only work if, in the case of natural selection, the offspring of a mating faithfully inherits the feature, or features, that confer(s) greater genetic fitness. What Darwin did not realize (nor for that matter did any other prominent contemporary biologist) was that while he was putting the finishing touches to the Origin of Species, the genetic basis of variation and the essential rules of inheritance were being painstakingly worked out in a monastery garden in Brno, in what is now the Czech Republic. The flowering of genetics The discipline of genetics was established on the basis of deductions made by Gregor (this was his Augustinian monastic name; his original forename was Johann) Mendel about the collection of artificially bred pea plants he maintained in the garden of his monastery. Mendel presented the results of his breeding experiments to the Natural Science Society in Brno in 1865, but he did not use the terms gene (meaning the smallest unit of heredity) or genetics. The word gene was not coined until 1909, nine years after Mendel's pioneering experiments came to the notice of evolutionary scientists. It was Mendel's good fortune that his various plant breeding experiments provided several examples of a simple one-to-one link between a gene and a trait – these are called single gene, or monogenic', effects. Mendel's simple dichotomies, yellow or green, smooth or wrinkled, are called discontinuous' variables.

In primate and hominin paleontology we normally have to deal with continuous' variables such as the size of a tooth, or the thickness of a limb bone. These have smooth, curved, distributions, not the neat columns that result from Mendel's data. How do you get continuous curves from discontinuous columns of data?

The answer is that many genes are involved in determining the size of a tooth, or the thickness of a limb bone, so that what looks like a curve is in reality the combination of many sets of columns.

10.8 Our Closest Relatives

Not so long ago a book on human origins would have devoted a substantial number of pages to descriptions of the fossil evidence for primate evolution. This was in part because it was assumed that at each stage of primate evolution one of the fossil primates would have been recognizable as the direct ancestor of modern humans. However, we now know that for various reasons many of these taxa are highly unlikely to be ancestral to living higher primates. Instead, this account will concentrate on what we know of the evolution and relationships of the great apes. It will review how long Western scientists have known about the great apes, and it will show how ideas about their relationships to each other, and to modern humans, have changed. It will also explore which of the living apes is most closely related to modern humans. Among the tales of exotic animals brought home by explorers and traders were descriptions of what we now know as the great apes, that is, chimpanzees and gorillas from Africa, and orangutans from Asia. Aristotle referred to apes' as well as to monkeys' and baboons' in his Historia animalium (literally the History of Animals'), but his apes' were the same as the apes' dissected by the early anatomists, which were short-tailed macaque monkeys from North Africa. One of the first people to undertake a systematic review of the differences between modern humans and the chimpanzee and gorilla was Thomas Henry Huxley. In an essay entitled On the relations of Man to the Lower Animals' that formed the central section of his 1863 book called Evidence as to Man's Place in Nature, he concluded the anatomical differences between modern humans and the chimpanzee and gorilla were less marked than the differences between the two African apes and the orangutan.

Darwin used this evidence in his —The Descent of Man‖ published in 1871 to suggest that, because the African apes were morphologically closer to modern humans than to the only great ape known from Asia, the ancestors of modern humans were more likely to be found in Africa than elsewhere.

This deduction played a critical role in pointing most researchers towards Africa as a likely place to find human ancestors. As we will see in the next chapter, those who considered the orangutan our closest relative looked to South-East Asia as the most likely place to find modern human ancestors. Developments in biochemistry and immunology during the first half of the 20th century allowed the search for evidence about the nature of the relationships between modern humans and the apes to be shifted from traditional morphology to the morphology of molecules. The earliest attempts to use proteins to determine primate relationships were made just after the turn of the century, but the first results of a new generation of analyses were reported in the early 1960s. The famous US biochemist Linus Pauling coined the name _molecular anthropology' for this area of research. Two reports, both published in 1963, provided crucial evidence. Emile Zuckerkandl, another pioneer molecular anthropologist, described how he used enzymes to break up the protein haemoglobin from blood red cells into its peptide components, and that when he separated them using a small electric current, the patterns made by the peptides from a modern human, a chimpanzee, and a gorilla were indistinguishable. The second contribution was by Morris Goodman, who has spent his life working on molecular anthropology, who used techniques borrowed from immunology to study samples of a serum (serum is what is left after blood has clotted) protein called albumin taken from modern humans, apes, and monkeys. He came to the conclusion that the albumins of modern humans and chimpanzees were so alike in their structure that you cannot tell them apart. Proteins are made up of a string of amino acids. In many instances one amino acid may be substituted for another without changing the function of the protein. In the 1960s and 1970s Vince Sarich and Allan Wilson, two Berkeley biochemists interested in primate and human evolution, exploited these minor variations in protein structure in order to determine the evolutionary history of the molecules, and therefore, presumably, the evolutionary history of the taxa being sampled. They, too, concluded that modern humans and the African apes were very closely related. Interrogating the genome, the discovery of the chemical structure of the DNA molecule meant that affinities between organisms could be pursued at the level of the genome.

This potentially eliminated the need to rely on morphology, be it traditional anatomy or the morphology of proteins, for information about relatedness. Now, instead of using proxies researchers can study relatedness by comparing DNA.

The DNA within the cell is located either within the nucleus as nuclear DNA, or within organelles called mitochondria in mtDNA. In DNA sequencing the base sequences of each animal are determined and then compared. Sequencing methods have been applied to living hominoids and the number of studies increases each year. The genomes of several modern humans and a few chimpanzees have been sequenced. Information from both nuclear and mtDNA suggest that modern humans and chimpanzees are more closely related to each other than either is to the gorilla. When these differences are calibrated using the best' palaeontological evidence for the split between the apes and the Old-World Monkeys, and if we assume that the DNA differences are neutral, the prediction is that the hypothetical ancestor of modern humans and the chimpanzee lived between 8 and 5 MYA. When other, older, calibrations are used, the predicted date for the split is somewhat older (e.g., >10 MYA).

Implications for interpreting the human fossil record The results of recent morphological analyses of both skeletal and dental anatomy, and the anatomy of the soft tissues such as muscles and nerves, are also consistent with the very strong DNA evidence that chimpanzees are closer to modern humans than they are to gorillas. But some attempts to use the type of traditional morphological evidence that is conventionally used to investigate relationships among fossil hominin taxa did not find a particularly close relationship between modern humans and chimpanzees. Instead, chimpanzees clustered with gorillas. This has important implications for researchers who investigate the relationships among hominin taxa.

They either need to use types of information about skulls, jaws, and teeth that are capable of confirming the close relationship between chimps and modern humans, or they need to find other sources of morphological evidence, such as information about the shape of the limb bones and see if those data are capable of recovering the relationships among living higher primates supported by the DNA evidence.

A traditional taxonomy and a modern taxonomy that take account of the molecular and genetic evidence that chimpanzees are more closely related to modern humans than they are to gorillas.

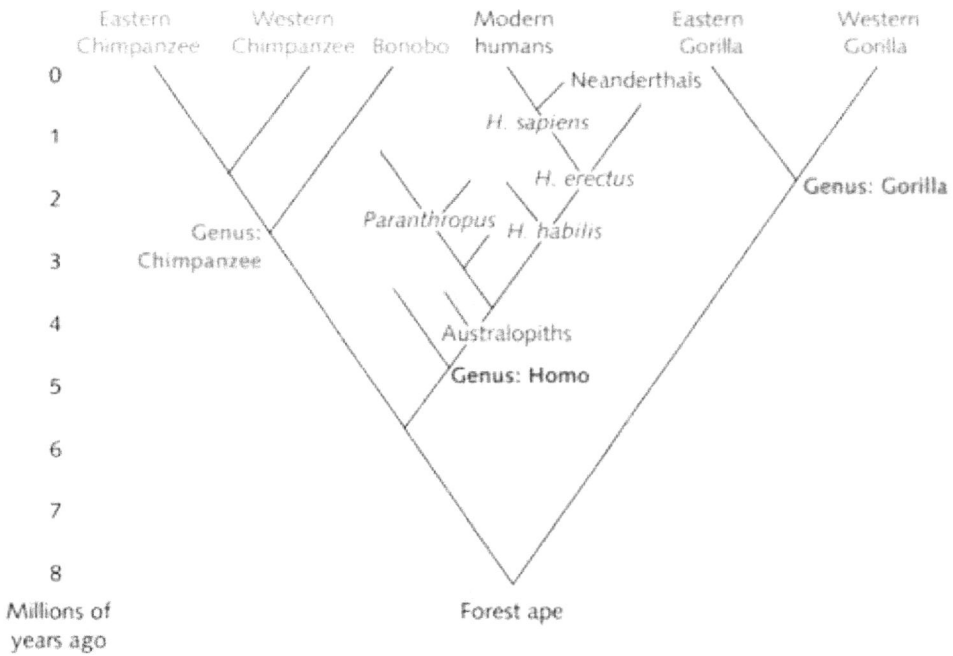

Figure 10-2 : Tree of life - Classification of Apes and Hominins into the hierarchical system

10.9 Fossil Hominins : Their Discovery and Context

As explained in previous chapter, a hominin is the label we give to anatomically modern humans and all the extinct species on, or connected to, the modern human twig of the Tree of Life. In this chapter I discuss what the hominin fossil record consists of, how it is discovered and how it and its context are investigated.

THE HOMININ FOSSILS RECORD

A fossil is a relic or trace of a former living organism. Only a tiny fraction of living organisms survives as fossils, and until people were buried deliberately, this also applied to hominins. We are almost certain that the fossils that do survive are a biased sample of the original population, and I discuss the implications of this in more detail in the next chapter. Fossils are usually, but not always, preserved in rocks. Scientists recognize two major categories of fossils. The smaller category, trace fossils, includes footprints, like the 3.6 MY-old footprints from Laetoli in Tanzania that I discuss in Chapter 6, and coprolites (fossilized faeces). The larger category, true fossils, consists of the actual remains of animals or plants. In the hominin fossil record they so outnumber trace fossils that when we use the word fossil it will normally apply to true fossils. Animal fossils usually consist of the hard tissues such as bones and teeth. This is because hard tissues are more resistant to being degraded than are soft tissues such as skin, muscle, or the gut.

> Soft tissues are only preserved in the later stages of the hominin fossil record: for example, the Bog People found in Denmark and elsewhere in Europe.

10.10 Fossilization

The chances that an early hominin's skeleton would have been preserved in the fossil record are very small. Carnivores, such as the predecessors of modern lions, leopards, and cheetahs, would most likely have had the first pick at the carcass of a dead hominin.

After them would have come the terrestrial scavengers, led by hyenas, wild dogs, and smaller cats, then birds of prey, then insects and finally bacteria. Within two to three years – a surprisingly short time – these organisms are capable of removing most traces of any large mammal. For its hard tissues to be preserved as fossils, the bones and teeth of a dead hominin would need to have been covered quickly by silt from a stream, by sand on a beach, or by soil washed into a cave. This protects the prospective fossil from further degradation and allows fossilization to take place. Fossilization of a bone begins when chemicals from the surrounding sediments replace the organic material in the hard tissues. Later on, chemicals begin to replace the inorganic material in bones and teeth. These replacement processes proceed for many years, and in this way a bone turns into a fossil. Fossils are essentially bone- or tooth shaped rocks. In the meantime, the sediments that surround the fossil are themselves being converted into rock. Teeth are already hard and durable in life, but chemical replacement also occurs in teeth. Diagenesis is the word scientists use to describe all the changes that occur to bones and teeth during fossilization. Fossils from different sites, and even fossils from different parts of the same site, show different degrees of fossilization because of small scale differences in their chemical environment. When fossils are preserved in hard rocks, and when they are freshly exposed, the fossils are very durable. However, if it is exposed to erosion by wind and rain for any length of time, fossil bone can be as fragile as wet tissue paper. In these cases, researchers have to infiltrate the fragile bone with liquid plastic, or its equivalent, in order to stop the fossil from disintegrating.

Obviously, deliberate burial greatly increases the chance that skeletons will be preserved in good condition. It is one of the main reasons why the human fossil record gets so much better about 60– 70 KYA. Most hominin fossils are found in rocks formed from sediments laid down by rivers, on lakeshores, or in one floors of caves.

Generally, older rocks (and thus the fossils they contain) are in the lower layers and the younger ones are nearer the surface: this principle is called the law of superposition. However, relative movement of rocks brought about by tension and compression, such as the shearing that occurs along faults in the earth's crust, can confound this general principle.

Sedimentary rocks that form in caves are also prone to be jumbled up in even more complex ways. Water that percolates down from the surface can soften and then dissolve old sediments. This produces Swiss-cheese-like cavities, which are then filled by more recent sediments. So, within caves, new sediments may be below old ones.

Earth scientists use the appearance, texture, and distinctive chemistry of rocks to describe and classify them. For example, they might refer to one layer as a pink tuff', or another as silty sand'. Just as there are rules for naming new species, there are rules and conventions for naming the strata of a newly discovered sedimentary sequence, and there is the equivalent of a Linnaean taxonomy for rocks. The layer of rock a fossil was buried in is referred to as its parent horizon'. Hominin fossils found within a particular rock layer are unless there is obvious evidence that they were deliberately buried, considered to be the same age as that layer. A fossil found embedded in a rock is described as being found in situ. Most hominin fossils, however, have been displaced through erosion from their parent horizon; these are called surface finds'. In order to reliably connect a surface, find to its parent horizon, it helps if the fossil still has some of the parent rock, or matrix, attached to or embedded in, it. This is why careful scientists never completely clean the matrix from a fossil.

10.11 Finding Fossil Hominins

Where do palaeoanthropologists look for early hominin fossils? In the 19th century Charles Darwin argued that, because the closest living relatives of modern humans, the chimpanzee, and the gorilla, were both confined to Africa then it was probable that the common ancestor of modern humans was also likely to have lived in Africa. So, for the past 75 years, and especially the last 50 years, Africa has been a focus of human origins field research.

But researchers cannot possibly search all of Africa. Are there particular places where hominin fossils are likely to be found? Palaeoanthropologists look where rocks of the right age (say back to 10 MYA) have been exposed by natural erosion. Erosion occurs in places where the earth's crust has been buckled and cracked as large landmasses, called tectonic plates, are pushed together. The area between major cracks or faults, is forced downward, and the earth's crust on the outside of the major faults is thrust upwards. This is how the floor and walls of rift valleys are formed. The faults that define the sides of rift valleys are sometimes so deep that the liquid core of the earth escapes through them. When it is under very high pressure, the molten core escapes as in a volcanic eruption. (Wood)

CHAPTER 11

NEANDERTHAL

&

CRO-MAGNON

11.1 *Neanderthal Genome*

Recent advances in high-throughput DNA sequencing have provided initial glimpses of the nuclear genome of Neanderthals as well as other ancient mammals including cave bears and mammoths. In the 7 May 2010 issue of Science, an international team of researchers presents the draft sequence of the Neanderthal genome composed of over 3 billion nucleotides from three individuals. Because Neanderthals are much closer kin to us than are chimpanzees, which diverged from the human lineage 5 to 7 million years ago, matching Neanderthal DNA against our own has the potential to reveal genetic changes that help define who we are.

11.2 About Neanderthals

Neanderthals (Homo neanderthalensis) are currently believed to be our closest evolutionary relatives. Although some researchers once thought they were our immediate ancestors in Europe, most now agree that Neanderthals and modern humans most likely shared a common ancestor within the last 500,000 years, possibly in Africa. The morphological features typical of Neanderthals first appear in the European fossil record about 400,000 years ago, with bones of full-fledged Neanderthals showing up at least 130,000 years ago. They lived in Europe and western Asia, as far east as southern Siberia and as far south as the Middle East (see map), before disappearing from the fossil record about 30,000 years ago.

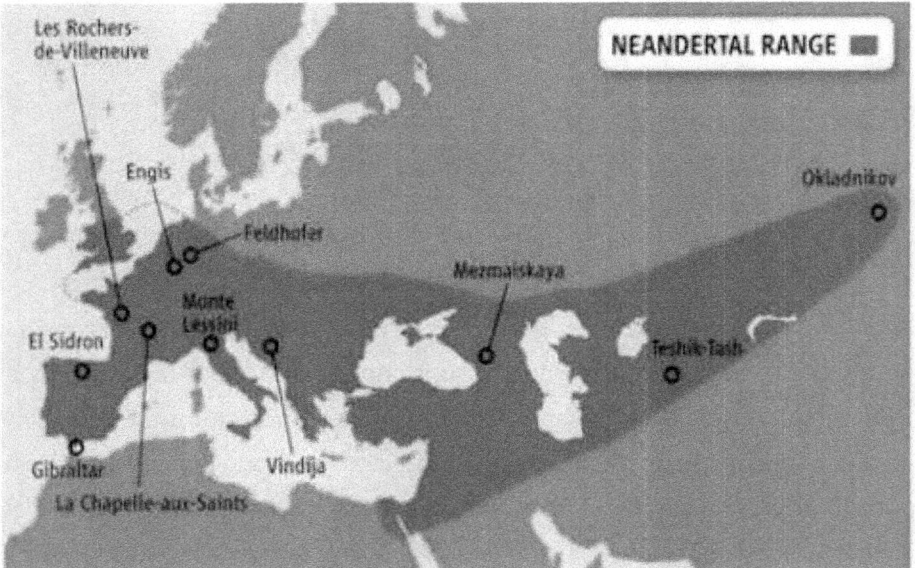

Figure 11-1 : Map of Eurasia - Neanderthals ranged from Europe to southern Siberia.

Fossil remains and anatomical reconstructions indicate that the typical Neanderthal had a stocky muscular body with short forearms and legs, a large head with bony brow ridges and a brain slightly larger than ours, a jutting face with a large nose, and perhaps reddish hair and fair skin. Neanderthals made and used a diverse set of sophisticated tools, controlled fire, organized their living spaces, hunted, and fed on game of various sizes, and occasionally made symbolic or ornamental objects.

The first Neanderthal fossils were discovered in 1829 in Engis, Belgium, and in 1848 at Forbes' Quarry, Gibraltar, but were not recognized as an early human species until after the 1856 discovery of "Neanderthal 1"-- a 40,000-year-old specimen, including a skullcap and various bones, found at the Kleine Feldhofer Grotte in the Neander Valley near Düsseldorf, Germany. This timeline highlights key discoveries about our closest relatives, from early fossil finds to the publication of the draft nuclear genome sequence.

Europe and the Middle East

A note to avoid confusion: It is a common misconception that Neanderthal and Cro-Magnon were the forerunners of Modern Man. Actually, Modern Man is much older than both of them. Accordingly, Modern man and the Humanoids are presented here, in the correct chronological order.

Originally Neanderthal and Cro-Magnon were not classed as Sapien (Wise). This designation was reserved for "us" Modern man. However, subsequent re-thinking by some - perhaps with other than scientific agendas, caused these two early Humanoids to be reclassed as Sapien. Thus, they became Homo-sapien neanderthalensis, and Homo-sapien Cro-Magnonensis, that leaves Modern Man with the really strange name "Homo-sapien-sapien" (Man the Wise Wise?). Though we do understand that the elevation of Cro-magnon and Neanderthal to Sapien status, does solve a great many problems - none related to science though. However, the current move to place Cro-magnon with modern man as "Homosapien-sapien", is really taking it too far, and it is pointless, as science proves, Caucasians did not evolve from Cro-Magnons in Europe.

Note: *The following images are various depictions of Neanderthals. Neanderthals may have varied in their appearance from region to region.*

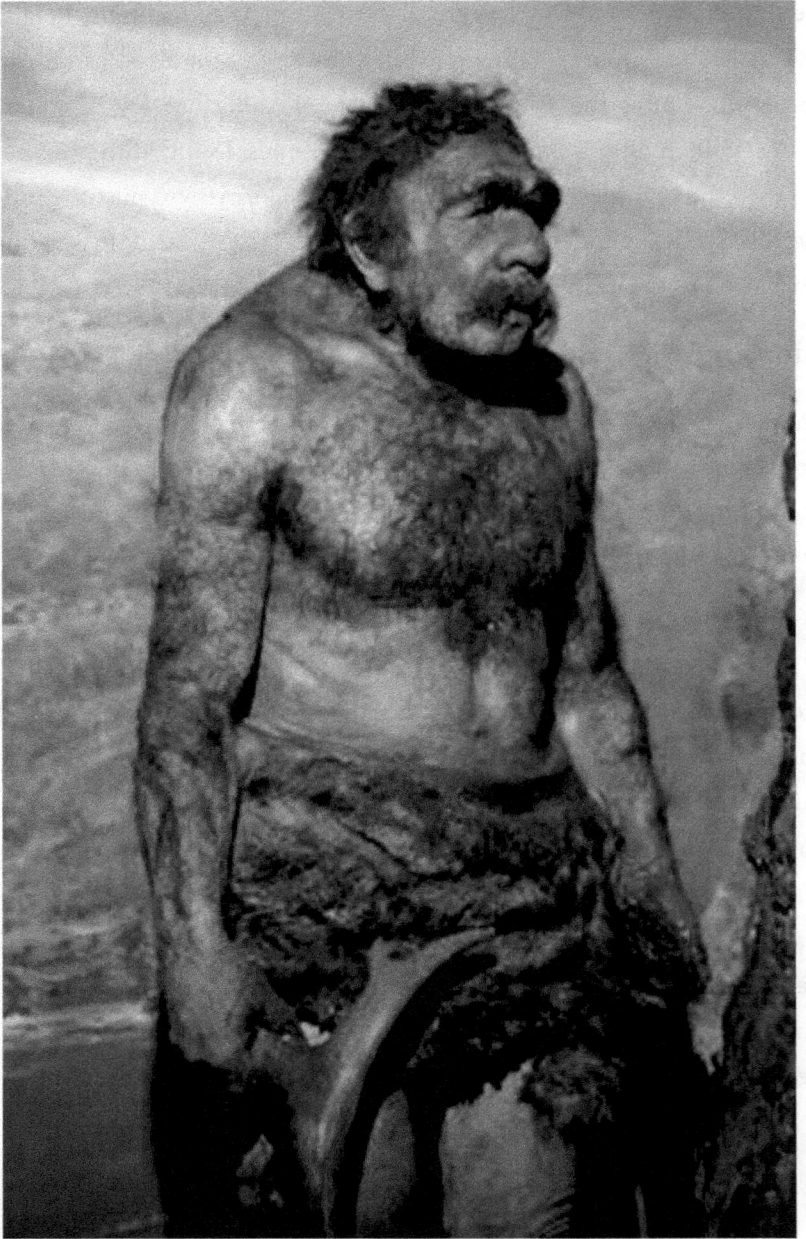

Figure 11-2 : A Neanderthal Hunter

196

Figure 11-3 (a & b) : A Neanderthal Family in their Cave Dwelling

Figure 11-4 : A Neanderthal Male Adult

Figure 11-5 : A Neanderthal Male Hunter

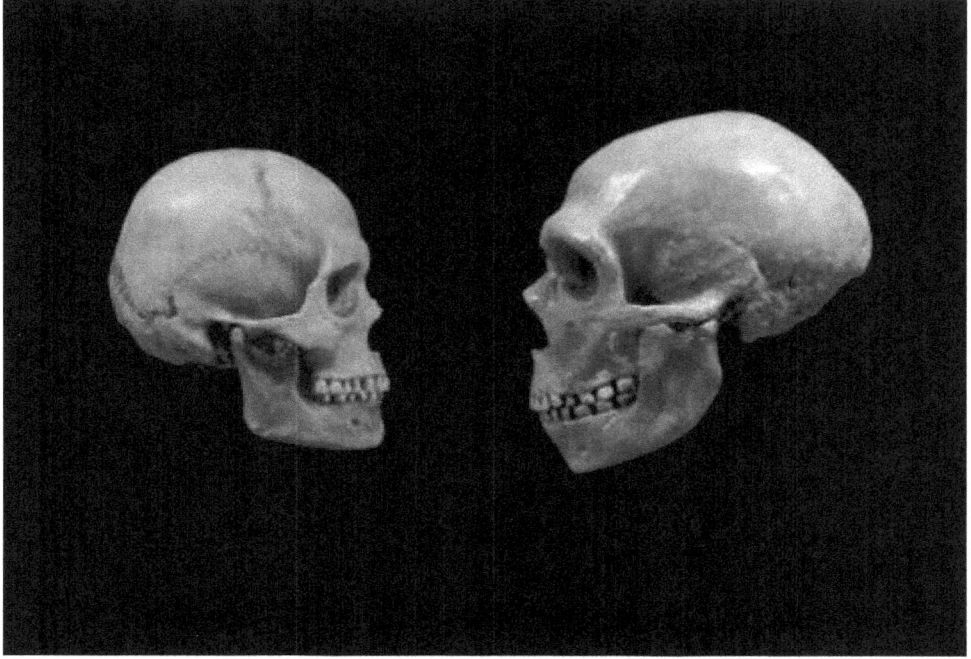

Figure 11-6 : Skull Comparison - A Modern Human Skull (left)
and a Neanderthal skull (right)

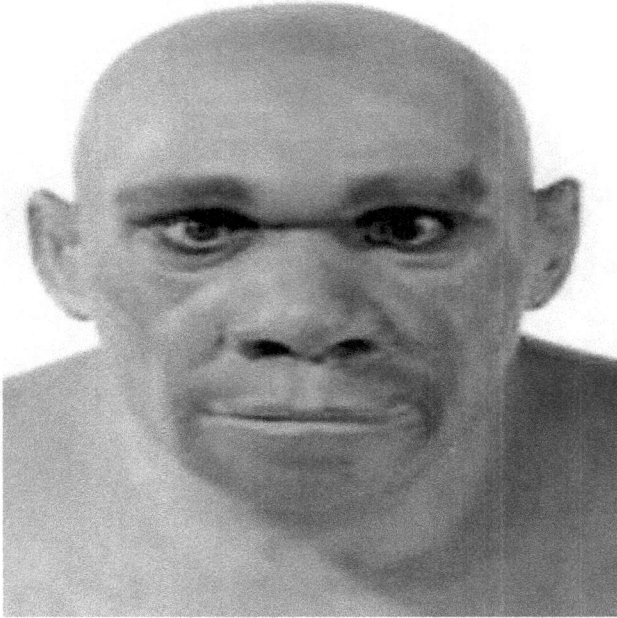

Figure 11-7 : Head model of a Neanderthal Male

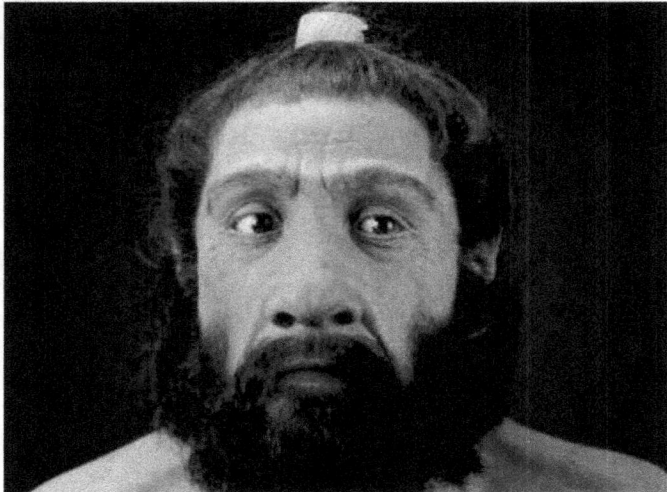

Figure 11-8 : Depiction of a Neanderthal Male

Figure 11-9 : A middle aged male Neanderthal Hunter

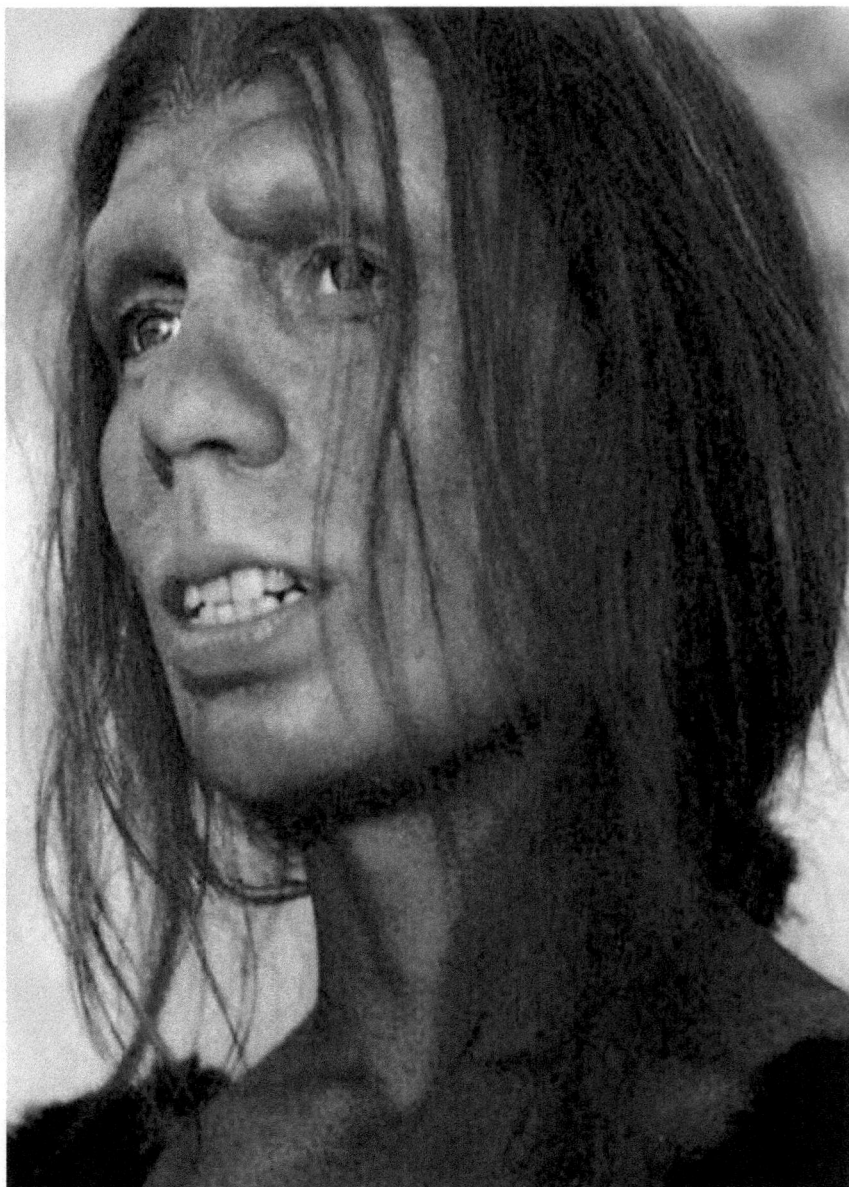

Figure 11-10 : A Neanderthal Female

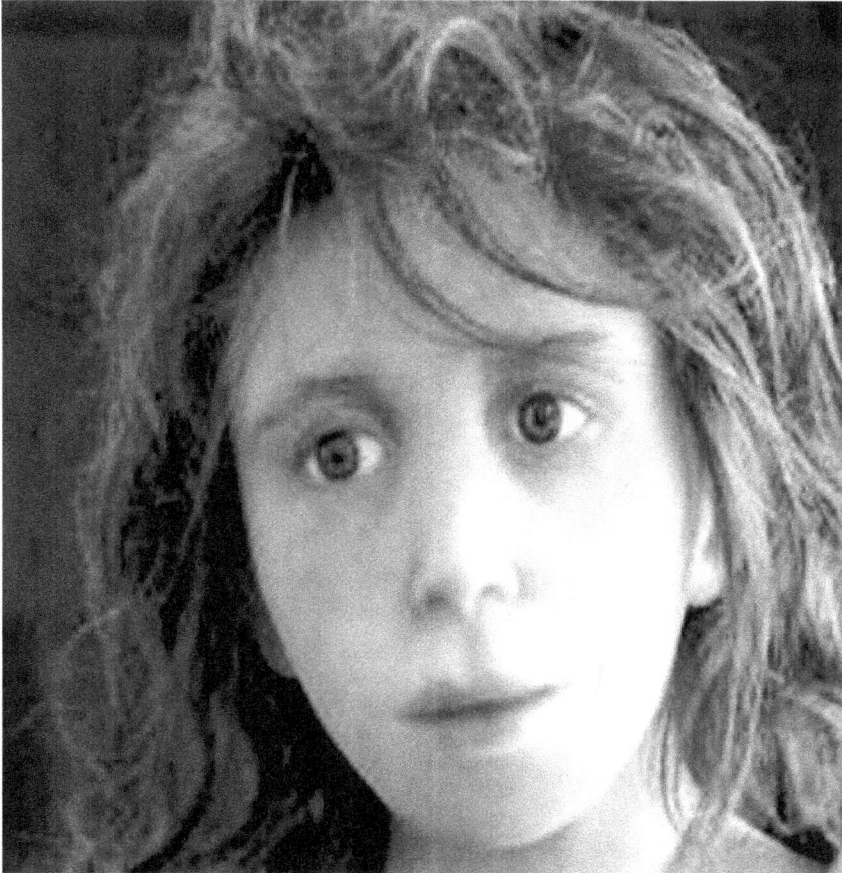

Figure 11-11 : Depiction of a Young Neanderthal Girl

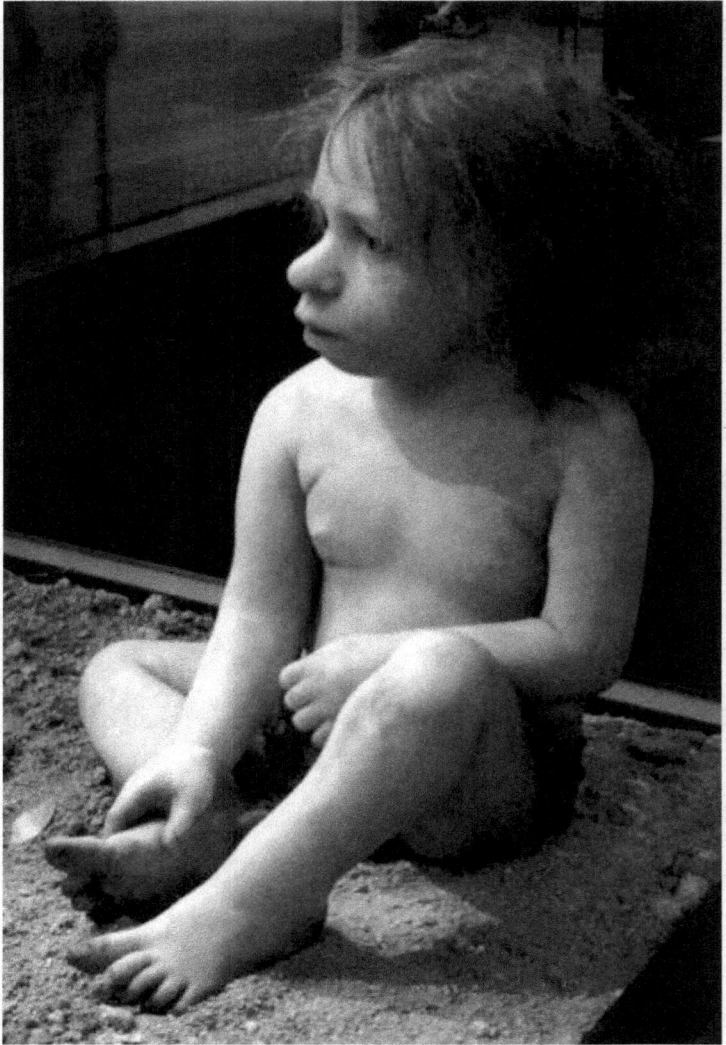

Figure 11-12 : Depiction of a young Neanderthal Child

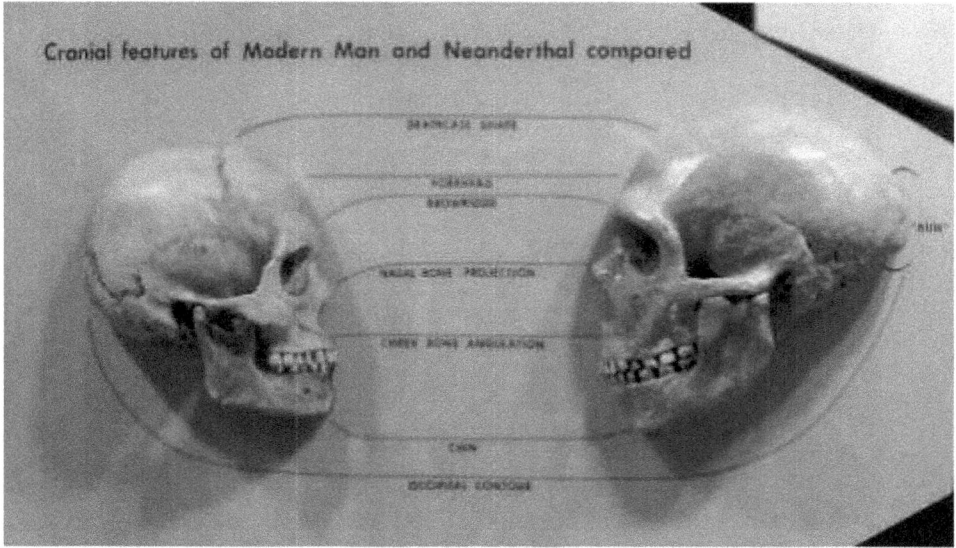

Figure 11-13 :Neanderthal skull having a slightly larger braincase when compared to modern human skull

Side by Side With an Ancient Relative

For the first time, a full Neanderthal skeleton has been assembled, using casts made from fossils. Here are key differences between it and modern humans.

	Neanderthal		Homo sapiens
BRAINCASE	Low		High
NASAL APERTURE	Large		Narrow
COLLARBONE	Long		Shorter
RIB CAGE	Cylindrical		Conical
LIMB BONES	Thick-walled		Thin-walled
TRUNK	Long		Short
HIP BONES	Flaring		Narrow
HAND BONES	Robust		Slender
JOINT SURFACES	Large		Smaller
LOWER LEG	Long		Short

Figure 11-14 : The Neanderthal skeleton is significantly shorter and wider when compared to a modern Human skeleton

11.5 Cro-Magnon

Cro-Magnon is the name of a rock shelter near Dordogne France. Here several prehistoric skeletons were found in 1868, these human remains are of the Upper Paleolithic period, 40,000 –10,000 years ago. Among these bones is the cranium and mandible of a male about 50 years old. This male is considered representative of the Cro-Magnon type, and this particular specimen is known as the —Old Man of Cro-Magnon."

The two figures below are skulls of Cro-Magnon. The Cro-Magnon skull is longheaded, the forehead is straight, the brow ridges only slightly projecting, the cranial vault noticeably flattened, and the occipital bone (at the back of the head) projects backward. The cranial capacity is large, about 100 cubic inches. Although the skull is relatively long and narrow, the face appears quite short and wide.

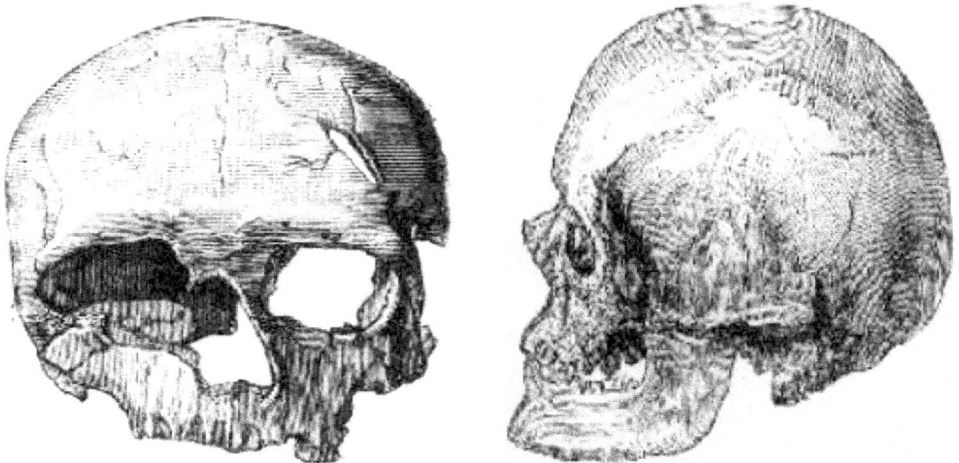

Figures 11-15 & 11-16 : Female and Male Cro-Magnon Skulls from left to right from Stuttgart found in 1884

This combination is often regarded as a common feature of the Cro-Magnon race. Cro-Magnon man is the name given to the oldest fossils found in Europe, which are anatomically identical to modern humans. They were discovered in a limestone cave in France in 1868. Four adult skeletons and one infant were found buried together along with pendants and necklaces made of pieces of shells and animal teeth, leading researchers to conclude that they were intentionally buried within the shelter.

The forward projection of the upper jaw is also distinctive. The eye sockets are low-set, wide, and rather square in shape; and the nasal aperture of the skull is narrow and strongly projecting. The mandible is robust, with massive ascending ramus (the upward projection of the lower jaw, where it attaches to the skull), has strongly developed points of muscular attachment, and a quite prominent chin. The stature of Cro-Magnon is from five feet five inches to five feet seven inches. Though in some areas they are taller.

The question of the relation of Cro-Magnons to the earliest forms of Homo-sapiens (like Neanderthal) is still unclear. It does appear however, that Cro-Magnons and Neanderthals are closer in affinity than was once believed. Though Cro-Magnon is found all over Europe, Asia and the Mediterranean, the tendency now is to locate the origin of the Cro-Magnon type Humanoid in the Middleeast: as typified by the remains found at the Jebel Qafzeh and Skhul sites in what is now Israel. Though the inescapable logical conclusion, is that Cro-Magnon is the product of Modern man crossbreeding with Neanderthal.

Like all the other theory's relating to early man, it has not yet been proven. But just as complex as the origin of Cro-Magnons, is the duration of Cro-Magnons. It appears that they only flourished during the Upper Paleolithic (old stone age 40,000 - 4,000 years ago).

11.6 Cro-Magnoids

Modern type individuals with at least some Cro-Magnon characteristics (these are called Cro-Magnoids), are found during the stone age in Europe, roughly from 5,000 to about 2,000 B.C. At the same time, remains have also been found for individuals who were quite different, often broad-headed, (as opposed to narrow headed).There are still some modern human groups that are thought to have retained a close relationship to Cro-Magnon types, at least in their cranial morphology. Particularly noteworthy of these are the Dal people from Dalecarlia (now Dalarna, Sweden.) and the Guanches of the Canary Islands, who are thought to represent a relatively pure Cro-Magnon stock.

Let us pursue the Guanches "Cro-Magnoids" a little further. These aboriginal peoples inhabited the western and eastern Canary Islands. They were first encountered by the conquering Spaniards at the beginning of the 15th century. Both populations are thought to have been of Cro-Magnon origin and may possibly have come from central and southern Europe via North Africa, in some distant age. Both aboriginal groups had brown complexion skin, blue or gray eyes, and blondish hair. These characteristics still persist in a large number of present-day inhabitants of the islands, but otherwise they are scarcely distinguishable in appearance or culture, from the current people of Spain. Neither original group now exists as a separate race.

When discovered by the Spaniards, these aborigines belonged to a stone age culture, though they were advanced enough to have pottery. Their food staples consisted mainly of milk, butter, goat flesh, pork, and some fruits; and their clothing was comprised of leather tunics or vests made of plaited rushes. They left alphabet like engravings and characters whose meanings are obscure.

11.7 A Closer look at Cro-Magnon and Neanderthal

Cro-Magnons and Neanderthals: this most classic of historical confrontations, sometimes couched in terms of brutish savagery versus human sophistication, has fascinated archaeologists for generations. On the one side stand primordial humans, endowed with great strength and courage, possessed of the simplest of clothing and weaponry. We speculate that they were incapable of fully articulate speech and had relatively limited intellectual powers. On the other are the Cro-Magnons, the first anatomically modern Europeans, with fully modern brains and linguistic abilities, a penchant for innovation, and all the impressive cognitive skills of Homo sapiens. They harvested game large and small effortlessly with highly efficient weapons and enjoyed a complex, refined relationship with their environment, their prey, and the forces of the supernatural world. We know that the confrontation ended with the extinction of the Neanderthals, perhaps about thirty thousand years ago. But how it unfolded remains one of the most challenging and intriguing of all Ice Age mysteries. The Neanderthals appeared on the academic stage with the discovery of the brow ridged skull of what seemed to be a primitive human in Germany's Neander Valley in 1856. Seven years later, Thomas Henry Huxley's brilliant study of the cranium in his Man's Place in Nature compared the Neanderthal fossil with the skulls of humankind's primate relatives, chimpanzees, and gorillas. The thought of a human ancestry among the apes horrified many Victorians. Public opinion carved out a vast chasm between archaic humanity, epitomized by the Neander Valley skull, and the modern humans discovered in the Cro-Magnon rock shelter at Les Eyzies, in southwestern France, in 1868. The Neanderthals became primitive cave people armed with clubs, dragging their mates around by their long hair. Unfortunately, the stereotype persists to this day. Cutting-edge science paints a very different portrait of the Neanderthals. They were strong, agile people who thrived in a harsh, often extremely cold Europe, from the shores of the Atlantic deep into Eurasia, from the edges of the steppe to warmer, drier environments in the Near East. Neanderthal hunters stalked large, dangerous animals like bison, then killed them with heavy thrusting spears. (Fagan)

They didn't have the luxury of standing off at a distance and launching light spears at their prey. But, for all their strength and skill, they were no matches for the Cro-Magnon newcomers, who, science tells us, spread rapidly across Europe around forty-five thousand years ago. Their hunting territories were small; they were thin on the ground; the routine of their lives changed infinitesimally from one year to the next. When they arrived in their new homeland, the Cro-Magnons were us, members of a species with a completely unprecedented relationship with the world around them. Every Cro-Magnon family, every band, was drenched in symbolism, expressed in numerous ways. Well before thirty thousand years ago, Cro-Magnons were creating engravings and paintings on the walls of caves and rock shelters. They crafted subtle and beautiful carvings on bone and antler and kept records by incising intricate notations on bone plaques. We know that they used bone flutes at least thirty-five thousand years ago, and if they did this, they surely sang and danced in deep caves by firelight on winter evenings and at summer gatherings. Cro-Magnons ornamented their bodies and buried their dead with elaborate grave goods for use in an afterlife. No one doubts that Cro-Magnon symbolic expression somehow reflects their notion of their place in the natural world. But their perceived relationship to nature was poles apart from our own—they were hunter-gatherers and lived in a world that was unimaginably different from today's Europe. And their perceptions of the world, of existence, were radically different from, and infinitely more sophisticated than, those of the Neanderthals. Cro-Magnon briefly explores the ancestry of the Neanderthals and the world in which they lived, then tries to answer the question of questions: What did happen when Cro-Magnon confronted Neanderthal? Did the moderns slaughter the primordial humans on sight, or did they simply annex prime hunting territories and push their ancient occupants onto marginal lands, where they slowly perished? Or did the superior mental abilities, hunting weapons, and other artifacts of the Cro-Magnons give them the decisive advantage in an increasingly cold late Ice Age world? Do we know what kinds of contacts took place between Neanderthal and newcomer?

Did the two populations intermarry occasionally, trade with one another, even borrow hunting methods, technologies, and ideas from each other? The answers to these questions revolve as much around the Cro-Magnons as they do the Neanderthals. Despite a century and a half of increasingly sophisticated research, the first modern inhabitants of Europe remain a shadowy presence, defined more by their remarkable art traditions and thousands of stone artifacts than by the nature of their lives as hunters and foragers, defined by the Ice Age world in which they flourished. Cro-Magnon paints a portrait of these remarkable people fashioned on a far wider canvas than that of artifacts and cave paintings.

Cro-Magnon is a story of hunters and gatherers who lived a unique adventure, whose earliest ancestors almost became extinct in the face of a huge natural catastrophe over seventy thousand years ago. It is a tale of ordinary men and women going about the business of survival in unpredictable, often bitterly cold environments that required them to adapt constantly and opportunistically to short- and long-term climate changed. These people were like us in so many ways: they had the same powerful intelligence and imagination, the ability to innovate and improvise that is common to everyone now living on earth. But they dwelled in a very different world from ours, one where pre-modern people still lived the same way they had hunted and gathered for hundreds of thousands of years. The history of the Cro-Magnons is the story of a great journey that began over fifty thousand years ago in tropical Africa and continued after the end of the Ice Age some fifteen thousand years ago. Above all, it's a story of endless ingenuity and adaptability.

Thanks to multidisciplinary science, we now know a great deal more about late Ice Age climate than we did a generation ago. Much of the raw material for this narrative does indeed come from artifacts and food remains, from abandoned hunting camps and the stratified layers of caves and rock shelters. New generations of rock art studies not only in western Europe but all over the world have added new perceptions about the meaning of Cro-Magnon art on artifacts and cave walls. However, compared with even twenty years ago, our knowledge of Europe's first moderns have changed beyond recognition thanks to technology and the now well-known revolution in paleoclimatology—the study of ancient climate.

Another revolution, in molecular biology, has added mitochondrial DNA (passed down through the female line) and the Y chromosome (roughly the equivalent in men) to the researcher's armory. We now possess far more nuanced insights into Neanderthal and Cro-Magnon life, especially into the environments in which they lived. Humans have always lived in unpredictable environments, in a state of flux from year to year. Until recently, we thought of the last glaciation of the Ice Age as a continual deep freeze that locked Europe into a refrigerator-like state for over one hundred thousand years, until about fifteen thousand years ago. Thanks to ice cores, pollen grains, cave stalagmites, and other newly discovered indicators of ancient climate, we now know that the glaciation was far from a monolithic event. Rather, Europe's climate shifted dramatically from one millennium to the next, in a constant seesaw of colder and warmer events that often brought near modern climatic conditions to some areas. Old models assumed that Scandinavia was buried under huge ice sheets for all of the last glaciation. Now we know that this was the case only during the Last Glacial Maximum, about 21,500 to 18,000 years ago, when much of Europe was a polar desert.

Much of the time Europe was far warmer, indeed near temperate. What is fascinating about the world of the Neanderthals and the Cro-Magnons is that we now have just enough climatological information to look behind the scenes, as it were, to examine the undercurrents of climate that caused hunting bands to advance and retreat and that perhaps helped drive some Neanderthal groups into extinction. Cro-Magnon explores Ice Age societies both historically obscure and well known, not just within the narrow confines of Europe, but on a far wider canvas. The Cro-Magnons may have been Europeans, but they were comparative newcomers who arrived from elsewhere. We cannot understand them without journeying far from the familiar confines of Les Eyzies and the Cro-Magnon rock shelter. Thanks to mitochondrial DNA and Y chromosomes, we know that they were ultimately Africans. Rather startlingly, we also believe that humanity almost became extinct in the aftermath of a colossal explosion, when Mount Toba, on Sumatra, erupted into space about 73,500 years ago. Connecting the dots between dozens of archaeological sites is one of the exciting challenges facing the archaeologist of the future.

Many of them are little more than scatters of stone artifacts, which we have to link to ash falls, to climate records wrested from cave stalagmites to the fluctuations of the Sahara Desert, and to the harsh realities of a life lived in often arid or cold landscapes. All we have at the moment is a tentative framework, based on frequently inadequate data. But it is enough to allow us to peer at the late Ice Age world not from the outside, but from within, for the fundamental routines of hunting and foraging in arctic and tropical, semiarid environments remain much the same today as they were over twenty thousand years ago. There are only a few options for, say, hunting reindeer with spears, driving rabbits into nets, or trapping arctic foxes. We know of them from historic as well as still living hunter-gatherer societies, whose basic subsistence activities have changed little over the millennia. The story of the Neanderthals and the Cro-Magnons tells us much about how our forebears adapted to climatic crisis and sudden environmental change. Like us, they faced an uncertain future, and like us, they relied on uniquely human qualities of adaptiveness, ingenuity, and opportunism to carry them through an uncertain and challenging world. We have much to learn from the remote past described in these pages.

11.8 Momentous Encounters

They call them LÖWENMENSCH, —the Lion Man.‖ The ivory figurine stands tall, leaning ever so slightly forward, arms by his sides. His head is a lion's, mouth slightly open, ears pricked, the mane cascading down the back. But his arms are human, relaxed, marked with six or seven striations (see color plate 1). The feet are slightly apart, a hint of maleness between the legs. The Lion Man stands serene, gazing calmly into the distance, contemplating an infinite landscape, a realm far beyond the confines of the living world. He came into being over thirty-four thousand years ago, carved out of water-soaked mammoth tusk by one of our remote ancestors, a Cro-Magnon.1 The artist who created the Lion Man was just like us. He laughed and cried, loved and hated, was calculating and sometimes devious. She was a member of a small hunting band, one of a few thousand people living in what is now southern Germany amidst a tapestry of coniferous forests and open tundra.

Here, reindeer herds migrated north and south with the seasons. Great mammoths fed by icy streams; flocks of arctic ptarmigan croaked at water's edge. This was no Ice Age paradise. The Lion Man's creator lived in a world whose harsh realities included frequent hunger and savage winters. But it was also a realm of the mind's eye, peopled with vibrant animals and powerful supernatural forces, which formed symbolic partnerships between humans and beasts. Löwenmensch, with his leonine head and human limbs, bridged the chasm between the living and supernatural realms, the kingdom of the imagination. His maker drew on the awesome cognitive abilities we ourselves possess. Nimble and tall, the Cro-Magnons were identical anatomically and intellectually to modern humans. We know that their brains had an identical configuration to ours, that they were capable of articulate speech, just as we are. The ancestors of the anonymous creator of the Lion Man had arrived in their challenging homeland about ten thousand years earlier from warmer and drier environments far to the southeast, in southwestern Asia.

A new generation of radiocarbon dates tells us that the Cro-Magnons spread across Europe within a mere five thousand years. People moved constantly, responding to social needs and to intelligence about game, campsites, and water supplies. The distances across Europe from southwestern Asia seem enormous, but within a few generations, Cro-Magnon bands would have covered surprising expanses, especially in sparsely populated, often bitterly cold environments, where climatic conditions were constantly changing, often for a few years at a time, sometimes for several lifetimes, at other times seemingly permanently. It's easy to imagine population movements that spanned 250 miles (400 kilometers) within a generation or so. And wherever they settled, the Cro-Magnons encountered small bands of Neanderthals, the European indigenes, people with biological and cultural roots hundreds of thousands of years in the remote past.2 About fifteen thousand years later, by about thirty thousand years ago, in one of the stunning developments of history, the Neanderthals were extinct. These pages tell the story of the Cro-Magnons, beginning with their encounters with the primordial Neanderthals.

The complex relationship between Cro-Magnon and Neanderthal has fascinated scholars for generations, as if it were the subject of an epic paleoanthropological novel. How did they perceive one another? Did they interbreed, or did the newcomers slaughter Neanderthals on sight? Were archaic and modern humans close neighbors, or did the Cro-Magnons simply push the indigenes out of their ancient hunting territories into marginal landscapes? Did the vastly superior intellectual abilities of the moderns play a central role in driving the Neanderthals into extinction, or were climate changes and extreme cold the ultimate villains? Reality, as far as we can know it, was far from an epic adventure. This is a story of brief but momentous encounters, of people separated by profound incomprehension and misunderstanding.

It is also a tale not of great leaders or powerful warriors, but of ordinary Ice Age people rising to the challenge of surviving in brutal environments. What were the secrets of the Cro-Magnons' brilliant success? Was it their more-advanced technology, their hunting and foraging abilities, or brilliant innovation combined with opportunism? Or did their spiritual beliefs and complex relationship with the supernatural realm play a decisive role? The portrait of the Neanderthals and the Cro-Magnons and their world in these pages comes from cutting-edge multidisciplinary science and a growing knowledge of the dynamics of hunter-gatherer societies in every corner of the world. It's no exaggeration to say that the foundations of today's Europe were forged in the events of the late Ice Age, between about forty-five thousand and twelve thousand years ago. We must begin by introducing the Cro-Magnons. In today's parlance, they are technically anatomically modern humans (AMHs). But the word Cro-Magnon rolls off the tongue much better and is a far more satisfying label for the first Europeans, even if it is technically somewhat incorrect. The name dates back to 1868, when the railroad came to the sleepy village of Les Eyzies, in southwestern France. Workmen clearing land for the new station uncovered a small, totally buried rock shelter and some flint tools and animal bones near a rock prophetically called Cro-Magnon, —great cavity."

A young geologist, Louis Lartet, dug into the back of the shelter its discovery and unearthed five human skeletons, including the remains of a fetus and several adults, among them a woman who may have been killed by a blow to the head. The burials lay among a scatter of shell beads and ivory pendants. These were no Neanderthals with simple artifacts and no bodily decoration. The Cro-Magnon people had round heads and high foreheads and were identical to modern humans. Les Eyzies lies on the bank of the Vézère River in a valley where high limestone cliffs with caves and deep overhangs provided wonderful shelter for Ice Age visitors.

Louis Lartet's father, Édouard, had partnered with Henry Christy, a wealthy English banker, to dig into Les Eyzies' huge rock shelters in the early 1860s. They had uncovered flint artifacts, engraved harpoons, and numerous reindeer bones, but no human remains. The Cro-Magnon find proved that the makers of these artifacts were Homo sapiens, the remote ancestors of modern Europeans, who lived during the Ice Age, during a period somewhat fancifully called l'Âge du Renne, or the Reindeer Age, because of the numerous bones of these animals found in the rock shelters. Soon scholars were comparing them (wrongly) to the Eskimo of the Arctic, but one fact was beyond question: they were the successors of the Neanderthals. Just where they came from is still the subject of lively academic debate. The Cro-Magnons, among whom the creator of the Lion Man numbered, were but specks on a vast European landscape of deep river valleys, mountains, and boundless open plains. They were well aware they were not the only humans preying on bison and reindeer, seizing meat from predator kills, stalking wild oxen on the edges of dark green pine forests. Just occasionally, they would glimpse their rivals— a Neanderthal band slipping quietly across a water meadow, people so different that Cro-Magnon children would run away. Like the Neanderthals, the Cro-Magnon newcomers were thin on the ground. But they were completely different. They were Homo sapiens, —the wise person,‖ capable of flexible thinking, planning ahead, and fully articulate speech. Europe was never the same after their arrival.

On long winter nights, the older men, perhaps those with unusual supernatural powers, would tell stories of a time long ago when their exotic neighbors were thicker on the ground. But even then, they were a rare presence, glimpsed walking quietly among the trees or high above a valley on a steep hillside. Now there were far fewer of them. Close encounters were an unusual event, perhaps during a hunt, or when collecting honey in the summer. Perhaps two handfuls of men and boys out hunting would face off unexpectedly, spears in hand, watching closely for a threatening gesture. The physical contrast was dramatic: tall, slender Cro-Magnons; compact Neanderthals. The Cro-Magnons wore closefitting fur parkas, long pants, and waterproof boots.

Their potential adversaries were barefoot men of immense strength, their bodies draped in thick furs crudely joined with thongs. They carried heavy, fire-hardened spears and wooden clubs, nothing more, weaponry virtually identical to that carried by their remote ancestors' tens of thousands of years before. Each side would stare at the other. Perhaps a few gestures would ensue, universal to all humans: a smile, a proffered gift of a honeycomb, perhaps some quiet grunts. There was no shared language, perhaps not even a common body odor. After a few moments, Cro-Magnon and Neanderthal would likely go their separate ways. We can only guess at the nature of such encounters. Our only potential analogies come from meetings between Western explorers and hitherto unknown societies, like, for example, the Tasmanian Aborigines, in the late eighteenth century. The Tasmanians had encountered no outsiders since rising sea levels had isolated them from mainland Australia nine thousand years earlier. Both sides recognized the other as fellow humans, but beyond that and some common gestures of friendship like a smile, they lived in entirely different worlds. We can be sure that any brief meetings with Neanderthals would be long remembered by the moderns, who would pass on recollections of such unusual events from one generation to the next. One is reminded of the New Zealand Maori, who still retained vivid memories of Captain James Cook and his ships a century after he departed over the horizon.

The Neanderthals, Homo neanderthalis, require no introduction to any reader of popular science. They are the cave people of prehistory, the hirsute folk with wooden clubs, a grossly unfair characterization of skilled, tough hunters armed with little more than wooden thrusting spears, who were not afraid to hunt such formidable beasts as the European bison and the aurochs, Bos primigenius, the fierce primordial wild ox. But there is much more to the Neanderthals than cartoonists' stereotypes. Forty-five thousand years ago, perhaps fifteen thousand to twenty thousand of them lived between the Atlantic Ocean in the west and the Ural Mountains, in Eurasia, far to the east. They hunted and foraged in small family bands. Most of them encountered no more than a few dozen fellow humans during their lifetime and then only briefly, perhaps for a cooperative hunt or to obtain a mate. They thrived for thousands of years in some of the most brutal environments.

11.9 Where Did They Come From?

What we know of their early history has been gleaned from meager scatters of stone tools and a few human fossils. These tell us that Neanderthal ancestry goes back far into remote prehistory, described in chapter 2. The earliest definite Neanderthals date to about 200,000 years ago. We also know, from an increasing number of archaeological sites, that their numbers rose slowly after about 150,000 years ago, following a period of intense cold that lasted at least 30,000 years. They flourished, albeit in smallish numbers, through subsequent, more temperate millennia. In what is now Italy, they hunted elephant and hippopotamus about 125,000 years ago. The warmer conditions lasted until about 115,000 years before the present, when the last glacial period of the Ice Age brought much colder temperatures and major environmental changes. By then, Neanderthals thrived in small numbers over an enormous area of Europe and Asia, from southern Britain and the Atlantic, through Belgium and France, across central Europe, and deep into central Asia, far east of the Black Sea into modern-day Uzbekistan and perhaps beyond. Neanderthal bands flourished in warmer environments, too, in the Near East in Greece, and in Spain as far south as Gibraltar.

For all their wide distribution, the small populations of the Neanderthals were a fleeting presence in grand Ice Age landscapes. They were the only human beings in a dangerous, predator-rich world, where survival depended on careful observation, constant watchfulness, and opportunism. Theirs was a rhythm of life that shifted infinitesimally over tens of thousands of years until the Cro-Magnons invaded ancient Neanderthal hunting territories and disturbed the even tenor of their days.

Cro-Magnons were the successors of the Neanderthals. But where had the Cro-Magnons come from? Had they originated in Europe itself, as many early Eurocentric scholars assumed? Or had they come from elsewhere? Experts pointed to the densely packed layers of rock shelters and caves in the Vézère Valley and northern Spain, to the seemingly orderly transition of artifacts from one stratified layer to another over thousands of years, starting with Neanderthal occupation and ending up at the end of the Ice Age.

Cro-Magnons were the successors of the Neanderthals. But where had the Cro-Magnons come from? Had they originated in Europe itself, as many early Eurocentric scholars assumed? Or had they come from elsewhere? Experts pointed to the densely packed layers of rock shelters and caves in the Vézère Valley and northern Spain, to the seemingly orderly transition of artifacts from one stratified layer to another over thousands of years, starting with Neanderthal occupation and ending up at the end of the Ice Age.

From a one-hundred-thousand-to eighty-thousand-year-old occupation level in one of the Klasies River caves, on the southeast African coast. These were the earliest Homo sapiens finds in the world at the time, and I, among others, had trouble accepting the chronology. At this point, molecular biologists studying mitochondrial DNA (mtDNA), which is inherited through the female line, threw an intellectual cat among the proverbial pigeons. A group of geneticists headed by Rebecca Cann and Alan Wilson, using mtDNA and a sophisticated —molecular clock,‖ traced modern-human ancestry back to isolated African populations dating to between two hundred thousand and one hundred thousand years ago. Inevitably there was talk of an —African Eve,‖ a first modern woman, the hypothetical ancestor of all modern humankind. Most archaeologists gulped and took a deep breath.

Cann and her colleagues had taken Homo sapiens into new and uncharted historical territory.5 Furious controversy surrounded the African Eve, pitting biological anthropologists who believed that all modern humanity had originated in Africa against those who argued for multiple origins in different parts of the Old World. As we will see in chapter 5, molecular biology is now much more refined, and the mtDNA, and now Y chromosome, samples are larger. The Genetic case for an African origin for Homo sapiens seems overwhelming. The archaeologists have also stepped forward with new fossil discoveries, including a robust 195,000-year-old modern human from Omo Kibish, in Ethiopia, and three 160,000- year-old Homo sapiens skulls from Herto, also in Ethiopia. Few anthropologists now doubt that Africa was the cradle of Homo sapiens and home to the remotest ancestors of the first modern Europeans—the Cro-Magnons.

The seemingly outrageous chronology of two decades ago is now accepted as historical reality. If Homo sapiens indeed originated in tropical Africa, how and when did the descendants of whom we can, somewhat indulgently, call the African Eve move into the semiarid lands of the Near East? Here we embark into the realm of speculation, largely because small bands of hunter-gatherers leave few traces of their passing behind them. We're back with the archaeological will-o-the-wisp, forced to rely on general clues. We can be pretty certain the migrants were people adapted to open country, who were constantly on the move. Their tools and weaponry had to be carried everywhere, so it is hardly surprising that little survives except for occasional groupings of stone tools. A generation ago, we thought of a single, albeit complex, movement out of Africa, perhaps about one hundred thousand years ago. This relatively simple model has given way to a more complex scenario involving two out-migrations. The first may indeed have occurred about one hundred thousand years ago but seems to have fizzled in the Near East, perhaps in the face of drought. A second, even less well-documented push seems to have taken place later, around fifty thousand years ago. This time, moderns settled throughout Near East Asia and stayed there, apparently living alongside a sparse Neanderthal population. This widely accepted theory assumes that by this time the newcomers had all the intellectual capabilities of Homo sapiens. Just when and how they acquired them remains a major unsolved problem.

All we can say is that at some point between one hundred thousand and fifty thousand years ago, at a seminal yet still little-known moment in history, Homo sapiens developed the full battery of cognitive skills that we ourselves possess. After a surprisingly short time, perhaps a mere five thousand years, their descendants moved northward into Eurasia and Europe. Like most archaeologists, I have a profound distrust of theoretical scenarios without the sites and artifacts to back them up, but in the case of modern-human origins, we have to work with such tools in the absence of much hard data. As we shall see in chapter 5, the two out-migrations theory seems the most convincing working model for bringing the ancestors of the Cro-Magnons into their new homeland. Refinements, indeed, wholesale changes, are likely to descend on this model in future years, but one fundamental point is of great importance: in the final analysis, the first modern settlers of Eurasia and Europe were ultimately Africans. Many of their hunting practices, light weapons, and social institutions developed in semiarid lands south of the Sahara.

11.10 Desert In The Tropics

This ancestry had a profound influence on the ways they adapted their lives to a much colder world of climatic extremes. The encounters between Neanderthals and Cro-Magnons in Europe and Eurasia were an intricate social gavotte that played out over many centuries. There was nothing new about the dance, for the newcomers had met Neanderthals many times before. Their ancestors had lived alongside Neanderthals in the semiarid lands of southwest Asia. You can be sure that oral traditions of their dealings with what to them must have been somewhat alien beings passed down through the generations. The Neanderthals were exotic because they lacked a common tongue, if, indeed, they possessed fluent speech at all. By the time the first Cro-Magnons arrived in Europe, a huge intellectual and social chasm separated them from their neighbors. We will, of course, never know what they thought of the Neanderthals.

They may have respected their great strength and their stalking abilities, but I suspect the Cro-Magnons thought of the Neanderthals not as humans like themselves, but as something that resembled them but acted, smelled, and spoke entirely differently. Each may well have avoided the other, for they had nothing in common. Their encounters were likely mostly momentary contacts in sheltered river valleys and on open plains, by lakes and on seacoasts, by rock shelters and in caves. Archaic and modern lived alongside one another, probably at a distance, until the last Neanderthals died out, probably in Spain, some thirty thousand years ago; the date is controversial. A huge academic literature surrounds the Neanderthals and their fateful encounters with the aggressive and opportunistic newcomers. Theories abound, as do questions, most of them virtually impossible to answer. Did modern humans attack their new neighbors and rapidly drive them out of their favorite hunting territories into marginal environments? To prove this would require dozens of regional maps with site distributions of numerous accurately dated Neanderthal and Cro-Magnon sites. Unfortunately, we don't have the sites, let alone a way of dating them accurately to within the span of a few generations, which is the kind of precision one would need. Take another scenario. Did Cro-Magnon bands kill off Neanderthals whenever they encountered them? Once again, the proof is near impossible to acquire. You would need to find human skeletons with spear points embedded in them— not just a single burial but dozens of them in different locations. So far, we have none. Then there's sex, a thorny subject that provokes news headlines without fail. Did Neanderthals and Cro-Magnons interbreed? Some years ago, the geneticists Svante Pääbo and Matthias Krings of the University of Munich succeeded in extracting a partial DNA sequence from a Neanderthal limb bone. Recently, Pääbo and his colleagues decoded most of the Neanderthal's mtDNA.6 They found that the Neanderthal sequence falls outside the range of Genetic variation in modern humans, which means they were not direct ancestors of Homo sapiens. Humans and chimpanzees share over 98 percent of their DNA sequence. Neanderthals were even closer to moderns, but the small differences are enough to show that we began to diverge from them around seven hundred thousand years ago. Could Cro-Magnons and Neanderthals then have interbred?

Most experts think they did not. There remains the most popular theory: modern humans were simply more adept at hunting and survival in very challenging, ever-changing late Ice Age environments. You can point to the Cro-Magnons' superior weapons and more efficient technology, to their clothing, and, above all, to their enhanced cognitive abilities. All of these must have been players in the ongoing gavotte, but to invoke a single, overriding cause for the Neanderthals' demise is to court accusations of oversimplification. Here, however, we can make a stronger case, based on both archaeological finds and intelligent speculation. Both Neanderthals and Cro-Magnons coped effortlessly with abrupt climatic changes from near-temperate to extremely frigid conditions. How well, however, the Neanderthals were able to deal with deep snow cover and long months of subzero temperatures is a matter of ongoing debate. They lacked what was, perhaps, one of the most revolutionary inventions in history, and an inconspicuous one at that: the eyed needle, fashioned from a sliver of antler, bone, or ivory.

If their expertise with antler is any guide, the Cro-Magnons must have been adept woodworkers in the more temperate environments of southwestern Asia. When they moved north, they settled on a continent where antler and bone were potential replacements for wood, and where mammoth and other large animal bones had to be used as fuel in more treeless environments. With brilliant opportunism, they used small stone chisels to remove fine splinters from antler and bone, which they then ground and polished into slender needles. Carefully fashioned stone awls served as drills to make the holes for the thongs that served as thread, substitutes for the vegetable fibers used with wooden needles in their ancestral homes. Every Cro-Magnon man, woman, and child must have been aware that protection from clothing came in layers, that warmth escaped from the head and extremities. As we will see, an indirect source of information on the garments they wore is the traditional clothing used by Eskimo and Inuit in very cold environments— the argument being that there are only a limited number of ways in which layered, cold-weather clothing can be fashioned from hides and skins.

The needle allowed women to tailor garments from the fur and skin of different animals, such as wolves, reindeer, and arctic foxes, taking full advantage of each hide or pelt's unique qualities to reduce the dangers of frostbite and hypothermia in environments of rapidly changing extremes.7 We cannot overestimate the importance of tailored clothing in Cro-Magnon life, especially when stacked up against draped skins. Cro-Magnon hunters also relied heavily on more lethal, lighter-weight stone-tipped spears with greater range, more effective weaponry than the fire-hardened weapons of their neighbors. In the long run, two innovations, layered clothing, and more effective projectiles, gave the Cro-Magnons a decisive practical advantage over the Neanderthals. The one enabled them to hunt efficiently in extremely cold, changeable conditions. The other allowed them to harvest a wider range of animals, especially medium-and smaller-sized beasts, which provided not only meat but also furs and other vital commodities for survival.

They also used ingenious devices for procuring small mammals, birds (including waterfowl in the late spring and summer), and eventually fish: nets, traps, light throwing darts, and so forth. Cro-Magnon technological imagination opened up for them a whole new ecological niche that had been beyond the reach of their predecessors. Was there another decisive advantage? The answer is probably an emphatic yes. What gave the newcomers the real edge was their intellectual awareness and imagination, their ability not only to cooperate with others, as the Neanderthals did, but also to plan ahead and to think of their surroundings as a living, vibrant world. This they defined with art and ritual, ceremony, chant, and dance, which helped them ride out the punches of rapid climate change and brutal temperatures, occasional hunger, and catastrophic hunting accidents. Their imaginations, their rituals, gave them a far more important cushion against harsh environments than any technological devices. We know from their art that they looked at their world with more than practical eyes, through a lens of the intangible that changed constantly over the generations. It was this symbolism, these beliefs, as much as their technological innovations and layered clothing, that gave them the decisive advantage over their neighbors in the seesawlike climatic world of the late Ice Age.

There were more of them living in larger groups than there were Neanderthals, too, so there were more intense social interactions, much greater food gathering activity from an early age, and an ongoing culture of innovation that came from a growing sophistication of language, advances in technology, and a greater life expectancy. In a world where all knowledge passed orally from one generation to the next, this enhanced cultural buffer between the moderns and the harsh climate provided an extra, albeit sometimes fragile, layer of protection during the intense cold of the so-called Last Glacial Maximum, from 21,500 to 18,000 years ago. There remain many uncertainties and profound disagreements over these various theories, but what do we know? There are no certainties, just some possible realities, which are a matter of instinct and extrapolation rather than scientific fact.

I think we can safely assume that most contacts between the two groups were sporadic and of short duration. Each side would have known the other was there watching, competing for game, and wary of attack, sudden contact, or simply loud shouting. The Cro-Magnons may have considered the Neanderthals, with their heavy brows and massive shoulders, repulsive in appearance—and vice versa. The contacts must have been like shadowboxing, with occasional sightings, sometimes little more than a presence glimpsed briefly among the trees. I think we can assume that the two groups were intensely curious about each other, both as potential competitors and as fellow humans, however different, with their own arcane knowledge and unique skills. We can be certain, too, that an intellectual void separated Neanderthal and newcomer. The Neanderthals still lived in the same simple way as their remote ancestors, in an annual round that changed infinitesimally over hundreds of generations. They lacked fully articulate speech and had few of the intellectual abilities of the newcomers. We do not even know if they believed in an afterlife, although they did sometimes bury their dead. They communicated with gestures and sound in ways honed over hundreds of thousands of years, but with nothing like the sophistication and panache of Homo sapiens. The Neanderthals were indeed human beings, but they lacked the humanity of the Cro-Magnons.

Unlike the Neanderthals, the Cro-Magnons celebrated, and sought to understand, their world with the help of their thoughts and imaginations, in chant, oral tradition, and ritual. Throughout Europe and Eurasia, the two peoples lived alongside one another for many generations, not necessarily competitors or afraid of one another, but always careful to observe each other's doings. The Neanderthals would have been the more silent presence, usually on the margins, usually invisible. They would never have been out of the Cro-Magnon mind. We can imagine a bright Cro-Magnon hearth in the heart of an open campsite on the plains in high summer, skin tents pitched on the perimeter. The people are dancing to a drum and flute, their profiles shimmering in the long shadows of the hearth.

They are under the spell of the dance, oblivious to the small group of Neanderthals, who are watching silently and invisibly just outside the circle of tents and firelight. When the dance ends, they will slip away without a sound, yet subconsciously the Cro-Magnons know they are there. There must have been a form of silent modus vivendi between modern and premodern, a tolerance based on incomprehension, yet a realization that the one had something to offer the other . . .

How did the Neanderthals and Cro-Magnons make contact other than by accidental encounter? They lived on the same continent, but they inhabited very different worlds, the one, as had always been the case, defined by the fundamental rhythms of the seasons, the other by the same endless cycle, but by much more as well. Their worlds and daily routines may have varied dramatically, but no hunting society is ever completely self-sufficient. The Neanderthals had an ancient, and profound, knowledge of what at first was unknown territory to the newcomers. They would have known where to find such esoterica as honeycombs, maybe supplied bearskins or hides to their neighbors, receiving perforated bear teeth or some tools and game meat in exchange. Ideas and hunting intelligence may have percolated between Neanderthal and newcomer as they watched each other. In chapter 6, I describe how some Neanderthal bands in France attempted to copy artifacts made by their more sophisticated neighbors. Given that the only common mode of communication between the two may have been gestures or grunts, perhaps some form of ―silent trade‖ may have brought them together.

By the standards of later times, the gavotte between Neanderthals and Cro-Magnons was history in slow motion, involving tiny numbers of people on both sides. Above all, it was a series of accidental encounters. Like so many other major events in human history, the arrival of the Cro-Magnons in Europe and Eurasia was not a deliberate act, planned for months and executed with precision, as if an army of hunters were on the march. Rather, modern humans arrived in dribs and drabs in Eurasia and what is now the Ukraine, on the banks of the Danube River, and eventually in the sheltered valleys of southwestern France. They settled along Mediterranean shores and as far west as northern Spain. Their movements consumed many, albeit short, generations, part of the endless population shifts of hunter-gatherer groups exploiting large hunting territories where food supplies were patchy and often widely separated. The number of people involved was infinitesimal, but their biological and cultural advantages over the indigenous population were enormous. Thanks to remarkably precise radiocarbon dates, we know that the Cro-Magnon move into the heart of the Neanderthal world came about haphazardly about forty-five thousand years ago. Small family bands followed migrating game into seemingly uninhabited lands. They found sheltered places to winter and locations where nuts were plentiful during the brief autumn. All kinds of trivial events triggered the expansion, all of them part and parcel of daily life: Sons split off from their families; bands fractured and dispersed when men or women quarreled with their siblings; people married into neighboring bands; hunters died in violent accidents, and their families joined other groups. The ebb and flow of people, of groups, fellow kin, families, and individuals, never ceased, propelled by the realities of the foraging life and its profound flexibility. And time and time again, Cro-Magnon encountered Neanderthal, with inconspicuous but ultimately momentous consequences for history. As we shall see, the initial settlement of Europe and Eurasia took place at an opportune time, when the climate to the north and west of the eastern Mediterranean was briefly warmer than it had been for a long time. The process of colonization was seemingly rapid, perhaps occupying but some five thousand years.

But the Cro-Magnons' initial steps into a harsh and unknown world disrupted and eventually destroyed a form of human existence that had remained virtually[495-511]unchanged for more than two hundred thousand years, its ancestry even deeper in the remote past, where our story begins. In its place, the newcomers created entirely new Ice Age societies, in which we can discern a distant mirror of ourselves.

11.11 Neanderthal Ancestors

Where did the Neanderthals ultimately come from? The search begins not in Europe but in the vastness of the African savanna, where the first humans flourished 2.5 million years ago. Parts of Africa are little changed since those remote days. You never forget Tanzania's Serengeti Plain, where the horizon stretches to infinity and the pale blue sky arches high overhead. Here you witness the African savanna much as it was two million years ago. The Serengeti is an endless, gently rolling plain of short, brown grass, interspersed with umbrella-like acacia trees, whose thin leaves provide a tracery of inadequate shade. Shallow gullies dissect the grassland with muddy water holes where wildlife congregates at dusk. Harsh, unforgiving, and arid for months on end, this seemingly inhospitable world supports a rich and varied bestiary that has changed little since the time of the first humans. Here elephants wallow at water's edge and giraffes browse on thorny trees. Tens of thousands of wildebeests migrate across the plain, like a forest on the move. And if you look closely, you'll almost certainly see a pride of lions resting in the shade or lacerating the carcass of a recent antelope kill, while hyenas await their turn and vultures hover and wheel overhead. As the late Harvard biologist Stephen Jay Gould once memorably remarked, we humans are all descendants from the same African twig. And it was here, on the African savanna, that the story of Europe, and of the Neanderthals, ultimately began. Here the remote ancestors of the first European Neanderthals learned how to hunt big game, virtually their only weapons a long wooden spear and an opportunistic ability to cooperate with others.

They lived in a state of constant watchfulness, alert for the predators that lurked on every side, relying on nimbleness and ingenuity for protection, knowing that their animal enemies were wary of bristling weapons. Their success at the hunt changed history. Less than a million years later, a few handfuls of premodern people crossed the Sahara and western Asia and settled in Europe, taking with them hunting skills honed on the semiarid savanna. Why should we look to Africa? The easy answer is that this is where humanity evolved, but that does not tell us why small numbers of pre-modern people crossed the Sahara and entered the Mediterranean and European worlds. Climate changes may have played a role, for the story of pre-modern Europe began about two million years ago, when Africa became progressively drier. The climatic shift had actually begun some five million years earlier, as tropical forests slowly gave way to more open scrub savannas, where seasonal aridity favored grasses and shrubs and a loose scatter of trees. These extensive, more open, and seemingly unproductive landscapes in fact provided accessible, more nutritious, and palatable food for herbivores of all kinds. People pursued these beasts. And these herbivores in turn provided food for humans. As the savanna spread in the face of more arid conditions, so did populations of herbivores, among them elephants, antelope, zebras, and rhinoceroses. These mammals became the dominant large animals of Ice Age Africa, Europe, and other parts of the world. They thrived in dry, seasonal environments because they consumed massive amounts of vegetable tissue, which required complex digestive tracts and large bodies. Antelope and other herbivores also ingest fiber-digesting microorganisms that process fibrous tissue in their guts. The large bodies that many of them have reflect their very active digestive systems. As the paleontologist R. Dale Guthrie remarks of the African eland, a large antelope, —the complex digestive physiology required to convert a coarse bit of gray shrub into a healthy eland is a little like trying to make cheese from chalk. The humans of two million years ago were not herbivores. People lived, for the most part, in open country and solved the problem of survival on the savanna by consuming the herbivores, which had digested the flora. They collected many plant foods, not only grasses and fruit but also tubers.

However, meat was the most important staple in the harsh, seasonal environments of the more arid lands. While it may have been the ideal food, collecting it by scavenging from predator kills was fraught with danger. The story of Europe begins in Africa not only because the remote ancestors of the Neanderthals evolved there, but also because that was where people learned how to hunt large animals.

Who was this first hunter? The trend toward greater aridity coincided with the appearance of a new species of human, known to scientists as Homo ergaster. Very different from and much more archaic human ancestors than the much later Neanderthals, these people stood upright, were fleet of foot, and had basically modern limbs, an elongated head with a strong browridge, and a brain capacity about three quarters of that of modern humans. Homo ergaster's anatomy reflected a life that involved covering long distances in open country. The species was larger than their predecessors, Homo habilis, too, the males weighing as much as 130 pounds (60 kilograms). The newcomers matured more slowly than apes and enjoyed a longer childhood, as well as a greater life expectancy, perhaps in their early twenties, although a much shorter one than that of modern people. Most likely, they lived in relatively small social groups, which ranged over large distances. Highly mobile and very watchful, they constantly acquired information about food and water supplies over many square miles of open country. Homo ergaster soon colonized large areas of Africa. Like the herbivores that were their prey, the hunting bands migrated with changing vegetation zones and moved alongside predators like lions, leopards, and hyenas, animals with whom they shared the ability of opportunistic hunting. The Neanderthals were expert hunters, but when did hunting begin? Once again, the answer lies in Africa. Homo ergaster was an omnivore, completely accustomed to quite drastic environmental changes in the distribution of open grassland, forest, and semiarid terrain and the dietary shifts that went with them. Unlike their predecessors, these people were serious hunters and meat eaters— because they dwelled for the most part in open country, where meat was the dominant, though not, of course, the only, food source. We know this because the bones of numerous large mammals appear alongside stone butchering tools in some of the archaeological sites that document their wanderings, whereas none appear in sites that predate them.

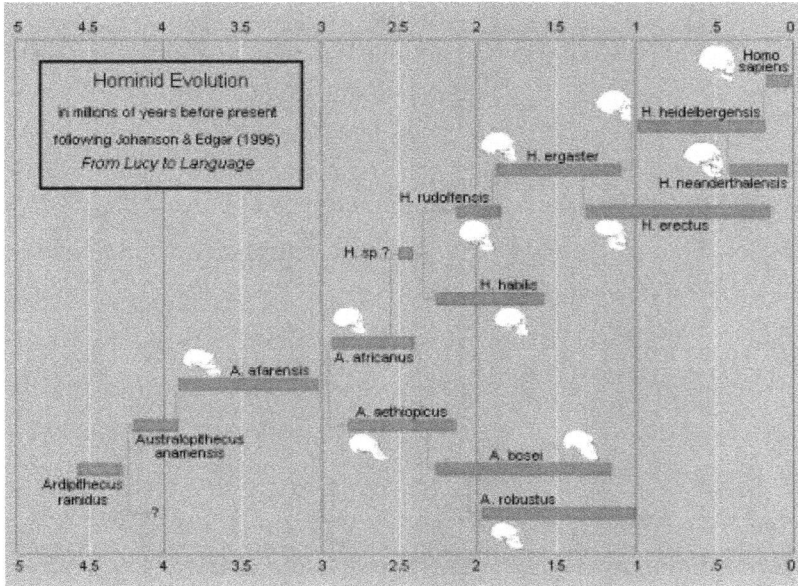

Figure 11-17 : A greatly simplified diagram of human evolution after two million years ago.

The Neanderthals were expert hunters, but when did hunting begin? Once again, the answer lies in Africa. Homo ergaster was an omnivore, completely accustomed to quite drastic environmental changes in the distribution of open grassland, forest, and semiarid terrain and the dietary shifts that went with them. Unlike their predecessors, these people were serious hunters and meat eaters – because they dwelled for the most part in open country, where meat was the dominant, though not, of course, the only food source.

The ancestry of both the Neanderthals and Homo sapiens lie with much earlier peoples. Living on the savanna was always hazardous, especially for relatively frail mammals whose only protections were their ingenuity and the simplest of weapons. Hunting even a medium-sized animal could be dangerous; predators awaited easy prey on every side. Antelope, zebras, and other animals protect themselves by being fleeter of foot than their predators, by kicking, or by using their horns.

Homo ergaster had no such defenses, nor any truly adequate security against large predators. No amount of shouting, stone throwing, or threatening behavior would deter a hungry lion, so every human band had to live extremely conservatively, with a constant eye for potential danger. Every group had young to be protected, for they matured slowly. They could not afford to become prey. The only potential defense came from weapons, objects that were, in effect, akin to sharp antelope horns. In their simplest forms, they were probably little more than thorny acacia branches or a long sapling with a tip sharpened with a stone flake. Half a dozen humans with such artifacts could readily deter a lion. As R. Dale Guthrie remarks, —even Hannibal's elephants bolted when faced with lethal rows of Roman pikes.‖3 The long, razor-sharp, and fire-hardened wooden spear developed over many generations from these simple beginnings to become one of the most significant, if rarely studied, weapons in human history. Not those weapons were everything. The humans' forms of hunting required extremely accurate perceptions of what was going on around them. Mammals, including humans, have disproportionately large brains that rely on intricate social relationships. So do carnivores, which hunt opportunistically. Brain size is largest among species that hunt large mammals opportunistically while cooperating with and depending on one another. Brain size also correlates with time spent as a juvenile, which in turn relates to exploration, learning, and play.

Complex social organization such as that possessed by Homo ergaster required intelligence gathering, analysis of that information, and creative uses of it. These hunting skills, and the weaponry that went with them, developed in Africa after two million years ago and survived virtually unchanged among premoderns everywhere for almost all of that time, until the late Ice Age, some fifty-five thousand years ago.

11.12 The Sahara Desert

1.8 million years ago, great flocks of birds gyrate above the shallow blue water, set in the midst of stunted grassland that undulates toward the sunbaked horizon. A gentle wind riffles the surface of the lake, where crocodiles lurk near water's edge.

As the shadows lengthen, great herds of antelope congregate there, drinking their fill as huge buffalo wallow in brown mud nearby. The plain shimmers in the late-afternoon heat, dark green and yellow after a recent shower. Lions yawn and stretch in the shade of a large boulder, lazily watching the lakeshore for a stray animal, for an easy kill. By another rocky outcrop, a small band of humans sit watchful, long wooden spears close to hand, ready for the evening hunt. On the far horizon, black clouds mass high above the distant mountains, whose sharp peaks stand out in front of the setting sun. As darkness falls, the wary hunters settle down on a high boulder, where lions will never catch them. The routine of life has not changed for generations—the unending cycle of dawn, midday, and dusk, of colder and warmer months, of dry season and wet. Like the bestiary of which the species is part, Homo ergaster shifts backward and forward across the at-this-time better-watered Saharan landscape. And when the desert dries up, and even before, tiny numbers of humans move out to its margins with the animals upon which they prey. Some return southward to the tropical savanna. Others move north and eastward, along the Nile Valley and Red Sea, and the Mediterranean coast. Their slow movements will change history. Around 1.8 million years ago—the date is still poorly defined— some groups crossed the Sahara Desert into western Asia. Human hunters were not just Africans. They radiated out of the continent as part of much larger mammalian communities that were to colonize Asia and Europe during brief periods of warmer climatic conditions in the north. Unfortunately, the finer details of climate change during most of the Ice Age are still little understood, despite generations of research. But we owe what we do know to the single-minded research of, of all people, a Serbian mathematician.

One of these briefs—but poorly researched— more temperate periods brought Homo ergaster and other mammals out of Africa and, ultimately, into Europe.

Not that this was a deliberate migration, for human societies lived by the rhythms of familiar routines that governed their daily existences. On an annual basis, Homo ergaster must have covered relatively short distances, maintaining close contact with other members of the social group, and perhaps neighboring bands. Everyone was accustomed to climatic shifts, to droughts and periods of higher rainfall.

The boundaries of where they could settle were set by their ability to reproduce and the availability of food and water. And, in the fullness of time, Homo ergaster (usually called Homo erectus outside of Africa) settled in western Asia and then in Europe (figure 2.2).

11.13 Moving Into Europe

Moving into Europe presented a challenge for humans, who were still basically tropical animals. Days were shorter for much of the year, winters were severe even during interglacials, and the seasons were very marked. Plant foods were in much shorter supply, which meant that game assumed even greater importance in human diet. One could not survive without an impressive expertise in big-game hunting.

Where and when, then, did pre-moderns first enter Europe? Most likely, the first Europeans entered their new homeland from western Asia across what is now Turkey, just as Homo sapiens would hundreds of thousands of years later. They lacked the watercraft to cross open water such as the Strait of Gibraltar. The immigrants were very thin on the ground, so it is hardly surprising that we know almost nothing about them. Some of them visited a small lake at Dmanisi, in Georgia's Caucasus Mountains, during a warm interval about 1.75 million years ago.7 Here the bones of Homo ergaster–like individuals and crude stone chopping tools were found under a medieval village. Unlike their African relatives, some of these people were short in stature. Their skulls display considerable anatomical variation, including traces of more primitive traits that predate ergaster in Africa. Interestingly, one individual had lost all but one of his teeth long before his death, for layers of bone fill his teeth sockets. He could only have consumed soft food that could be swallowed without chewing, so other members of the band must have looked after him. Dmanisi has strong African associations. The anatomy of the humans reflects their African ancestry. Fragmentary animal bones excavated from the site include those of short-necked giraffes and ostriches, both animals of African origin.

The people were part of a large bestiary, which also included such dangerous predators as saber-toothed cats. To survive must have required fleetness of foot and impressive hunting skills. Tiny numbers of human immigrants penetrated Europe during these millennia, so few that their bones rarely survive. When they do, the bones are often too fragmentary to tell us much.

In 2008, Spanish researchers discovered a 1.1-to 1.2-million-year-old fragmentary human jaw, as well as animal bones and stone tools, in the Sima del Elefante cave, in the Sierra de Atapuerca region of northern Spain, far to the west.8 We know that people were living at Ceprano, in central Italy, about 800,000 years ago, for a human skull found there resembles those of fossil humans from the Gran Dolina cave in the Sierra de Atapuerca dating to before 780,000 years ago. A few humans were in Britain by 700,000 years ago.9 A palimpsest of human bones, scatters of stone artifacts, and animal bones tell us virtually nothing about the first Europeans. One fascinating question immediately arises. Did they use fire? Could one survive in what were soon to become much colder environments without it? We don't know. Theories about the origins of fire abound, some of them placing its taming in tropical Africa at least 1.8 million years ago, but the evidence is little more than some fire-damaged stone that could have resulted from a wildfire. People may well have used fire from such sources thousands of years before they were able to light and control it, essential skills if it was to be used for cooking, protection, and warmth in cold climates.

The earliest known domesticated fire was made at a 790,000- year-old campsite at Gesher Benot Ya'aqov, in the Jordan Valley.10 Small numbers of people had already settled in temperate environments by this time, apparently without fire. But hearths lit at will provided warmth and, perhaps even more important, protection against predators, both in the open and in caves and rock shelters, so we can assume that fire was a decisive innovation as far as European settlement was concerned. Even with fire, however, much of glaciated Europe, with its savage winters, may have been too cold for human settlement. Like other mammals, the few groups in the north may have retreated southward into warmer areas during colder cycles.

It may be no coincidence that the earliest Europeans in central and western Europe appeared in Spain and Italy and that half a million years were to pass before human populations increased significantly during another period of warmer climate.

The Dmanisiskulls from Georgia display features reminiscent of their African ancestor, Homo ergaster, but was ergaster the direct ancestor of the Neanderthals? In 1907, workers in a sandpit at Mauer, near Heidelberg, Germany, unearthed a thick, chinless human jaw. Local experts promptly called it Homo heidelbergensis, —Heidelberg Man.‖ Few of their colleagues took the label seriously. They considered the incomplete fossil a European version of Homo erectus dating to about five hundred thousand years ago. Then, over the next three quarters of a century, a handful of fossils from Africa and throughout Europe revealed people with significant differences from Homo erectus who looked more like heidelbergensis. All had a higher, more filled-out braincase, reflecting a larger brain size closer to that of modern humans, a face that was reduced compared with the jutting countenances of earlier humans, and considerably thinner bony reinforcements of the skull. It turns out that the German anthropologists of a century ago were correct. Homo heidelbergensis was indeed a distinct human form with some more advanced features than erectus, but with more primitive anatomical characteristics than those of the Neanderthals and modern humans.11 Like Homo ergaster, heidelbergensis was African, and probably evolved there from earlier human stock about six hundred thousand years ago. Just when and how the newcomer moved out of Africa and into Europe remains a mystery, but the process must have been very similar to that of earlier humans— a natural movement alongside other mammals. Thanks to recent finds in Spain's Sierra de Atapuerca, we know that this little-known human was the direct ancestor of the Neanderthals.

Amphibian, lizard, and snake remains tell us the mean annual temperature was slightly warmer than today, averaging between fifty- and fifty-five-degrees Fahrenheit (ten and thirteen degrees Celsius).Exactly when the first true Neanderthals appeared remains unknown, for human fossils dating to between 250,000 and 70,000 years ago are scant. There are tantalizing hints: A Neanderthal-like skull from Ehringsdorf, near Weimar in central Germany, dates to perhaps about 200,000 years ago.

As the climate grew colder around 100,000 years ago, some Neanderthals camped around a mineral spring at Gánovce, near Poprad in Slovakia. At the time, coniferous trees were spreading into a region long covered with oak forests. Thereafter, Neanderthal remains become more common as intense cold settled over Europe and Eurasia.

Homo Heidelbergensis remains a shadowy presence: how did these people live, beyond apparently possessing fire? We can only generalize from a handful of well-documented sites. By half a million years ago, a few thousand hunter-gatherers dwelled throughout Europe, based for the most part in river valleys and other places where animals and plants abounded. Much of the time, they lived in small bands, with only sporadic contact with neighbors. Most people would have met only thirty to fifty people, if that many, during their brief lifetimes. The chances of hunting accidents or falling victim to predators were high for people possessing simple weaponry and living in an environment rife with large, dangerous animals and menacing predators like lions. For all that hazard, Homo heidelbergensis groups survived and flourished. For thousands of years, they lived by the banks of large rivers like the Somme, in northern France, and the Thames, preying on the game that populated these well-watered, sheltered locales, during millennia when climatic conditions were as warm as, if not warmer than, today.

> *Thousands of the Neanderthals could walk, and sense, their way through a late Ice Age landscape in ways that were soon to be compromised by full articulate speech. The only weapons for both men and women were their awareness, strength, stalking ability, and long hunting spears. They were quite people, who fed themselves by being inconspicuous and endlessly patient.*

The genetic bombshell came in a groundbreaking paper published by molecular biologists Rebecca Cann, Mark Stoneking, and Alan Wilson in Nature in 1987.

They presented the results of more than seven years spent collecting mitochondrial DNA from the placentas of newly born children. Their samples came from 147 individuals, whose ancestors lived in Africa, Asia, Europe, Australia, and New Guinea. After elaborate laboratory treatment, the samples yielded 133 distinct types of mtDNA. Some children had very similar sequences, as if they had descended from a single woman within the past few centuries. Others shared a common female ancestor, who had lived thousands of years age.

In the abstract to their, the three geneticists wrote, —All these mitochondrial DNAs stem from one woman, who is postulated to have lived around 200,000 years ago, probably in Africa.‖ Inevitably, science journalists labeled this shadowy ancestor —the African Eve.‖ She was, they said, our ten thousandth grandmother, when, of course, she was in fact part of a mall population.

A storm of controversy greeted the Nature paper. Biological anthropologists divided into two camps. One school of thought, known as multi-regionalists, argued that Homo sapiens had evolved from earlier humans in several regions of the Old World. Their opponents supported the geneticists and what became known as the —Out of Africa‖ hypothesis, the notion that Homo sapiens had originated in tropical Africa, then spread from there across the Ice Age world. A generation later, the furor has subsided in the face of new, even more sophisticated research involving both mtDNA and the Y chromosome. The genetic evidence is overwhelming. Homo sapiens evolved in Africa, from a common ancestral population that dates to somewhere around 170,000 years ago. As far as we can tell, no modern humans lived outside Africa until around 59,000 years ago.

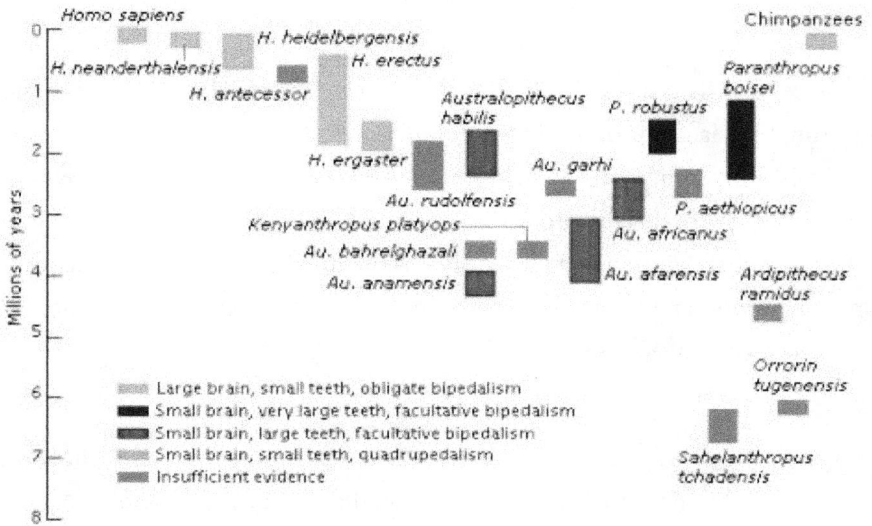

Figure 11-18 :Six million years of Transitional Human Fossils

11-14 Chasing Adam and Eve - Mitochondrial DNA and Y chromosomes

Since the 1980s, molecular biologists have studied modern-human origins, as well as the Genetic histories of modern populations. Two approaches have worked well. The first entails considering the genetic variations between living populations, the notion being to identify the most recent common ancestor of everyone living today. The second involves isolating DNA sequences from fossil Homo sapiens and pre-moderns such as Neanderthals.

Molecular researchers create phylogenetic trees based on data obtained from different parts of the genome and on the variability within these systems that one can observe in living societies. In most cases, they use variability in DNA sequences to create the family trees, the earliest common ancestor being the deepest point in time where the branches meet. Finding the primordial ancestor using phylogenetic trees is ultimately an exercise in statistical probability, based on assumptions about such factors as populations size and so on. It is one thing to identify the earliest starting point, and quite another to assign a date to him or her. This is where the researcher has to determine the rate of Genetic change and calibrate a molecular clock.

It should be stressed that phylogenetic trees give no indication of the behavioral or physical changes involved in the establishment of our species. That is where archaeology and skeletal anatomy come in. The genetic trees largely come from studies of mitochondrial DNA variation in living people. Mitochondria are the engines of cells, for they metabolize food and water into energy. They also maintain their distinctive DNA over tens of thousands of years. Mitochondrial DNA has only about sixteen thousand paired subunits of nucleotides (bases), is much easier to analyze than nuclear DNA, and has one priceless advantage. It is inherited only from the mother and passes intact from one generation to the next. This allows scientists to focus on the changes caused by mutations and mutations alone. By measuring the number of mutations that have taken place in the mtDNA of primates, whose evolutionary divergence millions of years ago has been dated from fossil bones, researchers have developed an mtDNA molecular clock.

Mitochondrial DNA mutates much faster than nuclear genetic material, changing every few hundred years. Thus, it can be used as a gauge of short-term evolution and especially as a timepiece for measuring when modern humans diverged from a common ancestor. Rebecca Cann and her colleagues published an evolutionary tree that showed that modern humans, Homo sapiens, had originated in Africa between 90,000 and 180,000 years ago.

More-recent analyses of the entire mtDNA genome by molecular biologist Michael Ingman and others have confirmed the findings of the original study. We now know that three of the deepest branches of the mtDNA tree are exclusively African, the next deepest being a mixture of Africans and non-Africans. All non-African DNA branches are of a very similar depth. Ingman and his colleagues believe that the mtDNA lineage evolved for some time in Africa, followed by an out-migration by a small number of people. A population bottleneck resulted, followed by a population expansion. All later European and Asian Homo sapiens lineages originated in this small African population. The researchers also refined the chronology, dating the most common recent ancestor to about 171,500 years (plus or minus 50,000 years) ago. The date of the earliest branch that includes both Africans and non-Africans is 52,000 years ago (plus or minus 27,500 years).

Y chromosomes are, in many respects, the male equivalent of mtDNA. The Y chromosome transmits across generations, but in the male. Much of it undergoes recombination, the rearrangement of genes that occurs during meiosis. The portions that do not are used to construct phylogenetic trees. In a landmark paper in 2000, twenty researchers studied a worldwide sample of Y chromosomes of men from dozens of populations on every continent. Using the same methods as were used in earlier mtDNA studies, they constructed a male family tree using the splits in the ancestry of the chromosome. The result was the same as that for mtDNA: the root of the tree lay in Africa. However, the —African

Adam lived not some 150,000 years but only 5,000 years in the past, more than 80,000 years after the African Eve. The chronology may prove too young.

Molecular biology provides a convincing general framework for the origins of Homo sapiens deep in Africa before 150,000 years age. The African Eve is a fictional person, a product of molecular biology, which has used mitochondrial DNA to show that all of us, wherever we live, are ultimately of African descent. If such a person existed, she would have been dark haired and black skinned, a member of small hunting band, and strong enough to tear apart human flesh with her hands and carry heavy loads. She was not the only woman on Earth, nor was she the most attractive, nor even the one with the most children. But she was so fruitful that a certain set of genes passed from her into every living being on Earth today. 33%.

When Bernard Vandermeersch excavated the Qafzeh cave, in Israel, he originally estimates that the modern-looking human burials from there were about forty thousand years old. A year after the publication of the Cann paper, French and Israeli scientists dated burned stone flakes from the site using the thermoluminescence method and came out with dates of about ninety-two thousand years ago. In one fell swoop, the chronology of Homo sapiens in the Near East moved back fifty thousand years. Recently, the Skhul burials were redated to virtually the same time period, much earlier than the original estimate of forty thousand years ago.

With both mtDNA and the new Qafzeh dates on the table, a new generation of theories argued that Homo sapiens had originated in tropical Africa some 150,000 years ago and had spread into the Near East by about 100,000 years before the present.

Why would such population movements have occurred? At the time when the Qafzeh people lived, goes the argument, the Near Eastern coast was essentially part of northeast Africa, better watered than today and a place that was readily accessible from both Arabia and the Nile Valley.

Molecular biology has traced the ancestry of the Cro-Magnons deep into tropical Africa, into the territory of the hypothetical African Eve. The genetic framework is plausible, but what can archaeology tell us to support it?

Without question, then, there were irregular human population movements across this vast, semiarid area about 100,000 years ago, perhaps through now dried-up and buried river valleys in the Sahara, which may have brought Homo sapiens groups into western Asia, either as occasional visitors or as permanent residents. There were many migrants; they probably hunter and foraged in almost the same way as their Neanderthal contemporaries, and their technology would have been virtually identical to that of their neighbors. In appearance, they would have displayed a mingling of more archaic and modern features, with the reduced brow ridges and modern vocal tracts of the Skhul people. Effectively, they were moderns, but their intellectual abilities were basically those of earlier humans, even if their stone tool manufacturing was a little more efficient and flexible than that of their predecessor.

What do we know about early Homo sapiens populations in Africa a the time of Qafzeh and before? Our information is sketchy at best. Judging from a handful of African fossils dating to between 300,000 and 50,000 years ago, the evolutionary process was a slow one that took as long as half a million years. During these millennia, Africans must have displayed a considerable range of variation, with, perhaps, the final—modernization of the skull and the enlargement of the brain, with its fully modern intellectual capabilities, occurring relatively late in the process. A few discoveries provide clues. A tall, well-built male with a broad forehead and thinning brow ridges lived at Omo Kibish, in Ethiopia, about 195,000 years ago. In 1997, the Ethiopian paleoanthropologist Yohannes Haile-Selassie unearthed three fully modern 160,000- year-old skulls, one of them from a child, at Herto, also in Ethiopia. Herto shows that by then the anatomical development of Homo sapiens had run its course, for the skulls are virtually identical to those of modern people.

Omo Kibish and Herto are conclusive proof that modern humans flourished in tropical Africa long before 100,000 years ago. Few in numbers, they lived in small, extremely isolated hunting bands. We know of this isolation from the National Geographic Society's Genographic Project, a major effort to track human migrations using DNA. As part of this ambitious research, a consortium of geneticist constructed a matrilineal family tree (passing through the female line) of 624 complete mtDNA genomes from living Africans.

They concentrated much of their effort on the Khoi and San peoples of southern Africa, because they are surviving representatives of ancient hunter-gatherer traditions – people with a slender, light build, quite different from the squat, cold-adapted build European Neanderthals. Their paternal and maternal lineages are along the deepest branches known among modern humans. The consortium believes that a major split in the human mtDNA tree occurred between 140,000 and 210,000 years ago, perhaps caused by genetic drift resulting from the per sis tent isolation of small human populations at the time. At this point, small Homo sapiens populations in East Africa and southern Africa became isolated from one another for about 70,000 years or so, until around 70,000 years ago.

11.15 Probing the Neanderthals

Now let us look at Neanderthal, he was an "in-between" form of Homo-sapien - less advanced than Modern man, but more advanced than Homo-Erectus. He inhabited much of Europe and the Mediterranean lands during the late Pleistocene Epoch, (about 100,000 to 30,000 years ago). Neanderthal remains have also been found in the Middle East, North Africa, and Central Asia. The name Neanderthal derives from the discovery in 1856 of the remains of this Humanoid in a cave above the Neander Valley in Germany, not far from Düsseldorf.

The last Glacial Ice stage in Europe was about 10,000 to 70,000 years ago, and it is from those times that the most numerous skeletal remains of Neanderthals have been found. These have given us some idea of Neanderthals body-type and habits. Neanderthals were short, stout, and powerful in build. Cranial capacity equaled or surpassed that of modern humans, though their braincases were long, low, and wide and flattened behind. Their faces had heavy brow ridges, large teeth, and small cheekbones. The chest was broad, and the limbs were heavy, with large feet and hands. The Neanderthals appear to have walked in a more irregular, side-to-side fashion than do modern humans.

Neanderthals were the first human group to survive in northern latitudes during the cold (glacial) phases of the Pleistocene. They had domesticated fire, as indicated by concentrations of charcoal and reddened earth in their sites. Yet, their hearths were simple and shallow and must have cooled off quickly, giving little warmth throughout the night. Not surprisingly, they exhibit anatomic adaptations to cold, especially in Europe, such as large body cores and relatively short limbs, which maximize heat production and minimize heat loss.

Neanderthals were cave dwellers, although they occasionally built camps out in the open. They wore clothing, used fire, hunted small and medium-sized animals (like goats and small deer), and they scavenged from the kills of large carnivores. They made and used a variety of stone tools and wooden spears. Neanderthals intentionally buried their dead, both individually and in groups, and they also cared for sick or injured individuals.

Evidence of ritualistic treatment of animals, which is sometimes found with their skeletons, may indicate that they practiced a primitive form of religion. Evidence from a few sites indicate that Neanderthals coexisted for several thousand years with Modern Humans, who arrived in Europe at about 45,000 B.C, and Cro-Magnons, who arrived in Europe by 35,000 B.C.

The origins of Neanderthals cannot be established with any certainty. The forerunners of Neanderthal humanoids may date to some 100,000 to 200,000 years ago. Some skull fragments found in France are of that age, but they have characteristics more like modern Homo sapiens. And so it may be, that this is where we see the first evidence of modern man (modern man first shows up at about 400,000 years ago, and is much older than Neanderthal); crossbreeding with Humanoids, in this case Homo-Erectus - who still existed as late as 300,000 B.C. Thus, producing the hybrid "Neanderthal".

Please take special note: The proclivity of all Humans to mate with whoever is available, is ingrained and pervasive. The activity of crossbreeding will eventually account for the introduction of all the modern world's races and ethnicities. It was thought that Neanderthals anatomic adaptations to cold had brought about a lightning of the skin, and thus played a part in the evolution of modern White people.

11.17 Findings

According to preliminary sequences, 99.7% of the base pairs of the modern human and Neanderthal genomes are identical, compared to humans sharing around 98.8% of base pairs with the chimpanzee. (Other studies concerning the commonality between chimps and humans have modified the commonality of 98% to a commonality of only 94%, showing that the genetic gap between humans and chimps is far larger than originally thought.)

The researchers recovered ancient DNA of Neanderthals by extracting the DNA from the femur bones of three 38,000-year-old female Neanderthal specimens from Vindija Cave, Croatia, and other bones found in Spain, Russia, and Germany. Only about half a gram of the bone samples (or 21 samples each 50-100 mg) was required for the sequencing, but the project faced many difficulties, including the contamination of the samples by the bacteria that had colonized the Neanderthal's body and humans who handled the bones at the excavation site and at the laboratory.

Additionally, in 2010, the announcement of the discovery and analysis of Mitochondrial DNA (mtDNA) from the Denisova hominin in Siberia revealed that this specimen differs from that of modern humans by 385 bases (nucleotides) in the mtDNA strand out of approximately 16,500, whereas the difference between modern humans and Neanderthals is around 202 bases. In contrast, the difference between chimpanzees and modern humans is approximately 1,462 mtDNA base pairs. Analysis of the specimen's nuclear DNA is under way and is expected to clarify whether the find is a distinct species. Even though the Denisova hominin's mtDNA lineage predates the divergence of modern humans and Neanderthals, coalescent theory does not preclude a more recent divergence date for her nuclear DNA.

11.18 History

In 2006, two research teams working on the same Neanderthal sample published their results, Richard Green and his team in Nature and Noonan et al. in Science. The results were received with some criticism, mainly surrounding the issue of a possible admixture of Neanderthals into the modern human genome. The speech-related gene FOXP2 with the same mutations as in modern humans was discovered in ancient DNA in the El Sidrón 1253 and 1351c specimens, suggesting Neanderthals might have shared some basic language capabilities with modern humans.

Svante Pääbo is the director of the Department of Genetics at the Max Planck Institute for Evolutionary Anthropology and head of its Neanderthal genome project. In 2006, Richard Green's team had used a then new sequencing technique developed by 454 Life Sciences that amplifies single molecules for characterization and obtained over a quarter of a million unique short sequences ("reads"). The technique delivers randomly located reads, so that sequences of interest, e.g., genes that differ between modern humans and Neanderthals, show up at random as well. However, this form of direct sequencing destroys the original sample so to obtain new reads more sample must be destructively sequenced.

Noonan et al., led by Edward Rubin, used a different technique, one in which the Neanderthal DNA is inserted into bacteria, which make multiple copies of a single fragment. They demonstrated that Neanderthal genomic sequences can be recovered using a metagenomic library-based approach.

All of the DNA in the sample is "immortalized" into metagenomic libraries. A DNA fragment is selected, then propagated in microbes. The Neanderthal DNA can be sequenced, or specific sequences can be studied. Overall, their results were remarkably similar. One group suggested there was a hint of mixing between human and Neanderthal genomes, while the other found none, but both teams recognized that the data set was not large enough to give a definitive answer.

The publication by Noonan et al. revealed Neanderthal DNA sequences matching chimpanzee DNA, but not modern human DNA, at multiple locations, thus enabling the first accurate calculation of the date of the most recent common ancestor of H. sapiens and H. neanderthalensis.

The research team estimates the most recent common ancestor of their H. neanderthalensis samples and their H. sapiens reference sequence lived 706,000 years ago (divergence time), estimating the separation of the human and Neanderthal ancestral populations to 370,000 years ago (split time).Earlier mitochondrial DNA research led by geneticist Svante Pääbo in 1997 had indicated present day Homo sapiens and Neanderthals mtDNA split into separate lineages approximately 500,000 years ago.

Green et al. calculated a divergence time of 516,000 years ago and do not indicate a split, while they claim the average divergence time between alleles within humans is thus 459,000 years with a 95% confidence interval between 419,000 and 498,000 years. These two dates (~500k) were calculated with assumption on nonselective pressure. If positive selection forced mtDNA changes then the split time may be shorter. In this study, the team stated: "Neanderthal genetic differences to humans must therefore be interpreted within the context of human diversity."

On the other hand, Noonan et al. found no evidence of Neanderthal admixture to the modern human genome, but they did not preclude admixture of up to 20% with a certainty better than 95%, and hence did not claim to present a definite answer to the question.

In February 2009, the Max Planck Institute's team led by Pääbo, announced that they had completed the first draft of the Neanderthal genome. An early analysis of the data suggested in "the genome of Neanderthals, a human species driven to extinction" "no significant trace of Neanderthal genes in modern humans".

New results suggested that some adult Neanderthals were lactose intolerant. On the question of potentially cloning a Neanderthal, Pääbo commented, "Starting from the DNA extracted from a fossil, it is and will remain impossible."

In May 2010, the project released a draft of their report on the sequenced Neanderthal genome. Contradicting the results discovered while examining mitochondrial DNA, they demonstrated a range of genetic contribution to non-African modern humans ranging from 1% to 4%. From their Homo sapiens samples in Eurasia (French, Han Chinese & Papuan) the authors state that it is likely that interbreeding occurred in the Levant before Homo sapiens migrated into Europe.

However, this finding is disputed because of the paucity of archeological evidence supporting their statement. The fossil evidence does not conclusively place Neanderthals and modern humans in close proximity at this time and place.

Previously, in 1999, a report was made of a rib fragment from the partial skeleton of a Neanderthal infant found in the Mezmaiskaya cave in the northwestern foothills of the Caucasus Mountains that was radiocarbon-dated to 29,195 ± 965 B.P., and therefore belonging to the latest lived Neanderthals. Ancient DNA was recovered for amtDNA sequence showing 3.48% divergence from that of the Feldhofer Neanderthal, some 2,500 km to the west in Germany. Phylogenetic analysis placed the two in a clade distinct from modern humans, suggesting that their mtDNA types have not contributed to the modern human mtDNA pool.

11.19 *Criticism*

A 2007 review of the data by Wall and Kim re-analyses the data obtained from the published papers of Noonan et al. and Green et al., and it holds that the results are inconsistent with each other. The review proposes serious problems with the data quality in one of the studies, possibly due to modern human DNA contaminants and/or a high rate of sequencing errors. The re-analyses confirmed both results to the Human-Neanderthal DNA Sequence Divergence Time (common ancestor), that is 706 kya (thousands of years ago) to the Noonan et al. analysis and 516 kya to the Green et al. analysis. The modern European-Neanderthal population split time was estimated at 35 kya for the Green et al. data, and 325 kya for the Noonan et al. data. Before, no split time was estimated by the Green et al. study, and according to Wall and Kim the split time originally estimated by Noonan et al. was even higher: 440 kya (the Noonan et al. paper mentions 370 kya).

While Noonan et al. were unable to definitively conclude that interbreeding between the two species of humans did not occur, they proclaim little likelihood of it having occurred at any appreciable level. The study opts for a 0% contribution of Neanderthal DNA to the modern European gene pool, based on the 95% confidence interval that indicates a margin between 0% and 20% contribution.

The re-analyses of Wall and Kim yielded interbreeding margins between 0% and 39% to the data of Noonan et al., and margins between 81% and 100% to the data of Green et al. These vastly inconsistent results could only be reconciled by assuming a very recent split time between the two populations of 60 kya or less. However, such a recent split time would not be consistent with the estimated modern European-Neanderthal population split time from the Noonan et al. data.

The key assumption of Noonan et al. is that the 38,000 years of fossilisation suffered by the Neanderthal DNA should have the genome analysis focus on ancient DNA fragments of about 50 to 70 base pairs in length.

Green et al. do not make such an assumption; they generalized towards the exclusion of modern human nuclear DNA contamination by finding little evidence of modern human DNA contamination. Such mitochondrial DNA tends to remain preserved longer than nuclear DNA. However, Wall and Kim noted a length dependence of the results, having the small fragments pointing to a divergence time similar to the results of Noonan et al. and the large fragments much more similar on average to modern human DNA – even to the extent of indicating an estimated human. Neanderthal sequence divergence time that is less than the estimated divergence time of two extant members of one referenced population in West Africa.

Although Wall and Kim hold modern human contamination to be size-biased, since Neanderthal DNA would be expected to have a tendency to be degraded into short fragments, they noted that length dependence of the results means that alignment issues alone are unlikely to be a sufficient explanation, since longer fragments would be easier to align and thus the data from longer fragments should be more accurate. Still, they mark this as a signal of potential contamination in the data of Green et al. No similar signal of potential contamination was found in the data of Noonan et al.

Contamination in the data of Green et al. should have decreased the Neanderthal-specific sequence divergence in this study. Since this is not the case, the assumption of contamination also would indicate a higher sequencing error rate in the Green et al. data since sequence errors would look the same as Neanderthal-specific mutations. These Neanderthal-specific mutations already were considered prone to error due to post-mortem DNA damage in both studies and were excluded from the results.

In summary, Wall and Kim consider a model with 78% contamination more likely than a model with no contamination and 94% admixture.

11.20 Neanderthal and Modern Human Interbreeding

The first genetic code of Neanderthal reveals interbreeding. The genetic code of the Neanderthal has been revealed for the first time, giving surprising clues to their intimate relations with modern humans, scientists report in Neanderthals are usually regarded as a separate species, Homo neanderthalensis. They were our closest relatives and they died out about 30,000 years ago.

An international team, including those at the Max Planck Institute in Germany, analyzed DNA from the remains of 3 Neanderthal individuals. They produced a sequence of the whole Neanderthal genetic code, or genome, the first time this has been done. They also compared the Neanderthal genome to modern humans, Homo sapiens, from different parts of the world.

11.21 Evidence for Interbreeding

Until now, scientists could only speculate whether Neanderthals ever interbred with modern humans, but the team's results revealed some surprises. They show that modern humans outside of Africa share genetic information with Neanderthals. This means modern humans probably interbred with Neanderthals soon after they left Africa around 60,000 years ago.

Professor Chris Stringer, the Natural History Museum's human origins expert, comments on the research and explains, 'This research suggests that the genomes of people from Europe, China and New Guinea lie slightly closer to the Neanderthal sequence than do those of Africans." The most likely explanation for this finding is that the ancestors of people in Europe, China and New Guinea interbred with Neanderthals (or at least with populations that had a component of Neanderthal genes) in North Africa, Arabia, or the Middle East, as they were exiting Africa, but before they spread out across the rest of the world .

Prof Stringer is one of the architects of the Out of Africa theory, which explains how all humans living today share an African origin and that those outside Africa migrated out in small groups during the last 60,000 years. His book The Origin of Our Species will be published early next year.

As well as comparing the Neanderthal genome with modern humans, the team also compared it with chimps. They found that genetic changes linked to skin and bone, metabolism, and brain functions, were unique to Homo sapiens.

11.22 *Previous Genetic Studies*

There have been genetic studies on Neanderthals before, as Prof Stringer explains. The first tiny piece of DNA from a Neanderthal fossil was published in 1997, and since then, with improvements in recovery techniques and computing power, 20 Neanderthals have yielded up increasing amounts of ancient DNA.' These DNA studies support evidence from the fossil record, showing that Neanderthals split from modern humans around 400,000 years ago. And similarly, studies on DNA from living people support fossil records showing that modern humans share an African origin within the last 200,000 years.

11.23 *A New Look at Neanderthal Relation*

Prof Stringer concludes, —As one of the architects of 'Out of Africa', I have regarded the Neanderthals as representing a separate lineage, and most likely a separate species from Homo sapiens." Although I have never ruled out the possibility of interbreeding, I have considered this to have been small and insignificant in the bigger picture of our evolution – for example, the results of isolated interbreeding events could easily have been lost in the intervening millennia. Now, the Neanderthal genome strongly suggests those genes were not lost, and that many of us outside of Africa have some Neanderthal inheritance.'

'Any functional significance of these shared genes remains to be determined, but that will certainly be a focus for the next stages of this fascinating area of research.'

11.24 Archaic Admixture

There is another hypothesis that anatomically modern humans interbred with Neanderthals during the Middle Paleolithic. In May 2010, the Neanderthal Genome Project presented genetic evidence that interbreeding did likely take place and that a small but significant portion of Neanderthal admixture is present in the DNA of modern Eurasians and Oceanians, and nearly absent in subSaharan African populations.

Between 4% and 6% of the genome of Melanesians (represented by the Papua New Guinean and Bougainville Islander) are thought to derive from Denisova hominins - a previously unknown species which shares a common origin with Neanderthals. It was possibly introduced during the early migration of the ancestors of Melanesians into Southeast Asia. This history of interaction suggests that Denisovans once ranged widely over eastern Asia. Thus, Melanesians emerge as the most archaic-admixed population, having Denisovan/Neanderthal-related admixture of ~8%.

In 2011, Jeff Walls from University of California studied whole sequence-genome data and found higher rates of introgression in Asians compared to Europeans. He also tested the hypothesis that contemporary African genomes have signatures of gene flow with archaic human ancestors and found evidence of archaic admixture in African genomes, suggesting that modest amounts of gene flow were widespread throughout time and space during the evolution of anatomically modern humans.

11.25 New DNA analysis shows ancient humans interbred with Denisovans

A new high-coverage DNA sequencing method reconstructs the full genome of Denisovans — relatives to both Neanderthals and humans — from genetic fragments in a single finger bone.(Katherine Harmon)

* The Denisova cave in Siberia was once home to a unique species of hominins.

Tens of thousands of years ago modern humans crossed paths with the group of hominins known as the Neanderthals. Researchers now think they also met another; less-known group called the Denisovans. The only trace that we have found, however, is a single finger bone and two teeth, but those fragments have been enough to cradle wisps of Denisovan DNA across thousands of years inside a Siberian cave. Now a team of scientists has been able to reconstruct their entire genome from these meager fragments. The analysis adds new twists to prevailing notions about archaic human history.

"Denisova is a big surprise," says John Hawks, a biological anthropologist at the University of Wisconsin–Madison who was not involved in the new research. On its own, a simple finger bone in a cave would have been assumed to belong to a human, Neanderthal or other hominin. But when researchers first sequenced a small section of DNA in 2010—a section that covered about 1.9 percent of the genome—they were able to tell that the specimen was neither. "It was the first time a new group of distinct humans was discovered" via genetic analysis rather than by anatomical description, said Svante Pääbo, a researcher at the Max Planck Institute (M.P.I.) for Evolutionary Anthropology in Germany, in a conference call with reporters.

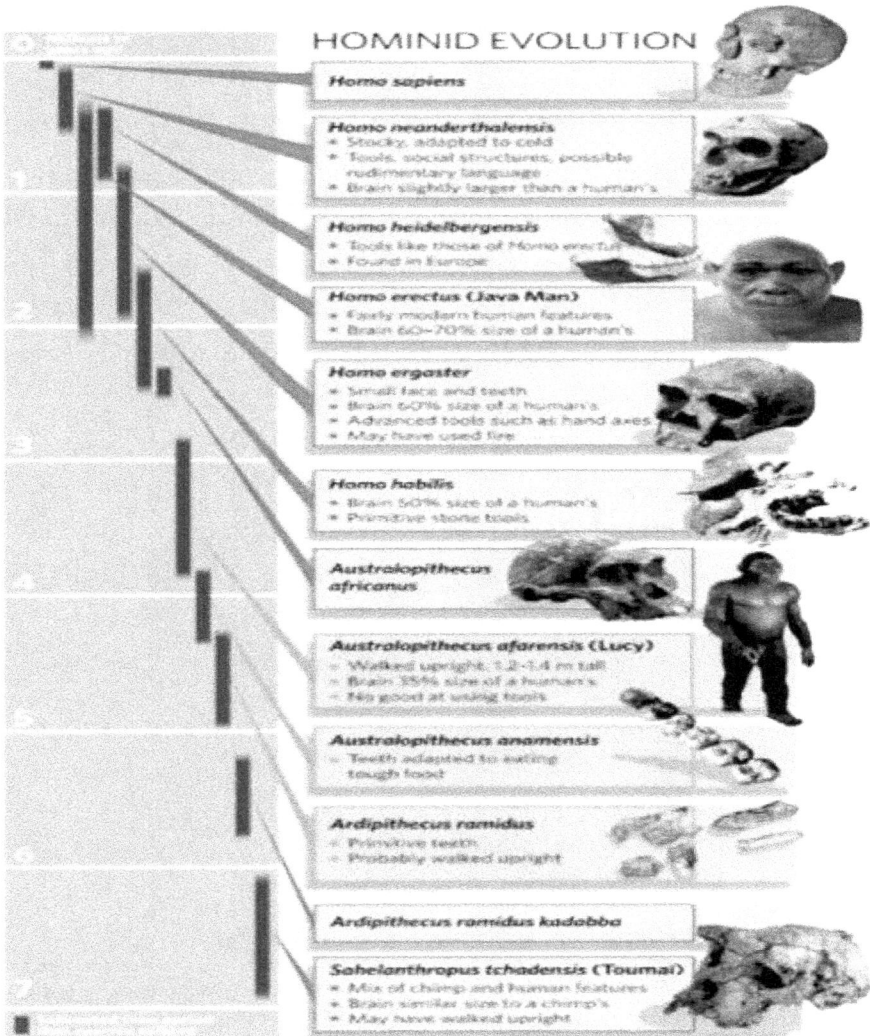

Figure 11-4 : Hominid Evolution - Common Ancestry

11.26 Fossil Genome Reveals Ancestral

Now Pääbo and his colleagues have devised a new method of genetic analysis that allowed them to reconstruct the entire Denisovan genome with nearly all of the genome sequenced approximately 30 times over akin to what we can do for modern humans. Within this genome, researchers have found clues into not only this group of mysterious hominins, but also our own evolutionary past. Denisovans appear to have been more closely related to Neandertals than to humans, but the evidence also suggests that Denisovans and humans interbred. The new analysis also suggests new ways that early humans may have spread across the globe. The findings were published online August 30 in Science1.

11.27 Who were the Denisovans

Unfortunately, the Denisovan genome doesn't provide many more clues about what this hominin looked like than a —pinky bone‖ does. The researchers will only conclude that Denisovans likely had dark skin. They also note that there are alleles "consistent" with those known to call for brown hair and brown eyes. Other than that, they cannot say.

Yet the new genetic analysis does support the hypothesis that Neanderthals and Denisovans were more closely related to one another than either was to modern humans. The analysis suggests that the modern human line diverged from what would become the Denisovan line as long as 700,000 years ago—but possibly as recently as 170,000 years ago.

Denisovans also interbred with ancient modern humans, according to Pääbo and his team. Even though the sole fossil specimen was found in the mountains of Siberia, contemporary humans from Melanesia (a region in the South Pacific) seem to be the most likely to harbor Denisovan DNA. The researchers estimate that some 6 percent of contemporary Papuans' genomes come from Denisovans. Australian aborigines and those from Southeast Asian islands also have traces of Denisovan DNA.

This suggests that the two groups might have crossed paths in central Asia and then the modern humans continued on to colonize the islands of Oceania.

Yet contemporary residents of mainland Asia do not seem to posses Denisovian traces in their DNA, a "very curious" fact, Hawks says. "We're looking at a very interesting population scenario"—one that does not jibe entirely with what we thought we knew about how waves modern human populations migrated into and through Asia and out to Oceania's islands. This new genetic evidence might indicate that perhaps an early wave of humans moved through Asia, mixed with Denisovans and then relocated to the islands—to be replaced in Asia by later waves of human migrants from Africa. "It's not totally obvious that that works really well with what we know about the diversity of Asians and Australians,"

Hawks says. But further genetic analysis and study should help to clarify these early migrations.

Just as with modern Homo sapiens, the genome of a single individual cannot tell us exactly what genes and traits are specific to all Denisovans. Yet just one genome can reveal the genetic diversity of an entire population. Each of our genomes contains information about generations far beyond those of our parents and grandparents, said David Reich, a researcher at the Massachusetts Institute of Technology–Harvard University Broad Institute and a co-author on the paper. Scientists can compare and contrast the set of genes on each chromosome—passed down from each parent— and extrapolate this process back through the generations. "You contain a multitude of ancestors within you," Reich said, borrowing from Walt Whitman.

The new research reveals that the Denisovans had low genetic diversity— just 26 to 33 percent of the genetic diversity of contemporary European or Asian populations. And for the Denisovans, the population on the whole seems to have been very small for hundreds of thousands of years, with relatively little genetic diversity throughout their history.

Curiously, the researchers noted in their paper, the Denisovan population shows "a drastic decline in size at the time when the modern human population began to expand."

Why were modern humans so successful whereas Denisovans (and Neanderthals) went extinct? Pääbo and his co-authors could not resist looking into the genetic factors that might be at work. Some of the key differences, they note, center around brain development and synaptic connectivity. "It makes sense that what pops up is connectivity in the brain," Pääbo noted. Neanderthals had a similar brain size–to-body ratio as we do, so rather than cranial capacity, it might have been underlying neurological differences that could explain why we flourished while they died out, he said.

11.28 More from Scientific American

Hawks counters that it might be a little early to begin drawing conclusions about human brain evolution from genetic comparisons with archaic relatives. Decoding the genetic map of the brain and cognition from a genome is still a long way off, he notes—unraveling skin color is still difficult enough given our current technologies and knowledge.

11.29 New Sequencing for Old DNA

The Denisovan results rely on a new method of genetic analysis developed by paper co-author Matthias Meyer, also of M.P.I. The procedure allows the researchers to sequence the full genome by using single strands of genetic material rather than the typical double strands required. The technique, which they are calling a single-stranded library preparation, involves stripping the genetic material down to individual strands to copy and avoids a purification step, which can lose precious genetic material.

The finger bone—just one distal phalanx—is so small that it does not contain enough usable carbon for dating, the researchers note. But by counting the number of genetic mutations in a genome and comparing them with other living relatives, such as modern humans and chimpanzees, given assumed rates of mutations since breaking with a last common ancestor, "for the first time you can try to estimate this number into a date and provide molecular dating of the fossil," Meyer said.

With the new resolution, the researchers estimate the age of the bone to 74,000 to 82,000 years ago. But that is a wide window, and previous archaeological estimates for the bone are a bit younger, ranging from 30,000 to 50,000 years old. These genetic estimations are also still in limbo because of ongoing debate about the average rate of genetic mutations over time, which could skew the age.

"Nevertheless," the researchers noted in their paper, "the results suggest that in the future it will be possible to determine dates of fossils based on genome sequences."

This new sequencing approach can be used for any DNA that is too fragmented to be read well through more traditional methods. Meyer noted that it could come in handy for analysis of both ancient DNA and contemporary forensic evidence, which also often contains only fragments of genetic material.

Hawks is excited about the new sequencing technology. It is also helpful to have a technology developed specifically for the evolutionary field, he notes. "We're always using the new techniques from other fields, and this is a case where the new technique is developed just for this."

Hawks himself has heard from the researchers that have worked with the Denisovan samples that "the Denisovan pinky is just extraordinary" in terms of the amount of DNA preserved in it. Most bone fragments would be expected to contain less than 5 percent of the individual's endogenous DNA, but this fortuitous finger had a surprising 70 percent, the researchers noted in the study. And many Neandertal fragments have been preserved in vastly different states— many are far worse off than this Denisovan finger bone.

The new sequencing approach could also improve our understanding of known specimens and the evolutionary landscape as a whole. "It's going to increase the yield from other fossils," Hawks notes. Many of the Neandertal specimens, for example, have only a small fraction of their genome sequenced. "If we can go from 2 percent to the whole genome, that opens up a lot more," Hawks says. "Going back further in time will be exciting," he notes, and this new technique should allow us to do that. "There's a huge race on—it's exciting."

11.29 New Sequencing for Old DNA

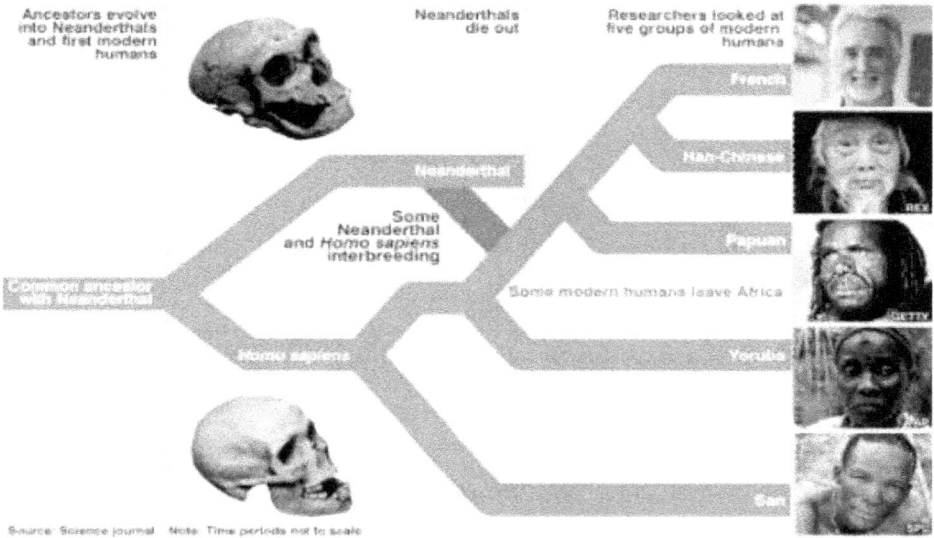

Figure 11-5 : Common Ancestry of Homo Sapiens & Neanderthal

The Denisovans might be the first non-Neanderthal archaic human to be sequenced, but they are likely not going to be the last. The researchers behind this new study are already at work using the new single-strand sequencing technique to reexamine older specimens. (Meyer said they were working on reassessing old samples but would not specify which specimens they were studying—the mysterious "hobbit" H. floresiensis would be a worthy candidate.) Pääbo suggests Asia as a particularly promising location to look for other Denisovan-like groups. "I would be surprised if there were not other groups to be found there in the future," he said.

Taking this technique to specimens from Africa is also likely to yield some exciting results, Hawks says. Africa, with its rich human evolutionary history, holds the greatest genetic diversity. The genomes of contemporary pygmy and hunter–gatherer tribes in Africa, for example, have roughly as many differences as do those of European modern humans and Neandertals. So "any ancient specimen that we find in Africa might be as different from us as Neandertals," Hawks says. "Anything we find from the right place might be another Denisovan."

11.30 Denisovans, Indigenous Australia's Siberian Kin

Indigenous Australians and other Melanesian peoples have evolved in part from a hitherto unknown strain of Neanderthal-like Siberians. German researchers using genetic material extracted from a 40,000-year-old finger bone and tooth discovered in a cave in Russia's Altai Mountains found their owners were part of a race of hominin distinct from both the ancestors of modern humans and the Neanderthals. They named them the Denisovans after the Denisova cave in which the bones were found.

Scientists believe Neanderthals and homo sapiens separated from a common ancestor about 500,000 years ago. The discovery of the Denisovans, reported in the journal Nature today, showed that Neanderthal-like people were not confined to Europe, said the professor of human evolution at the Australian National University, Colin Groves. "What it now suggests is that the European branch, the Neanderthaloids, were actually a full Eurasian branch, not just European -- and that they had an eastern and a western subdivision," he said.

Furthermore, by comparing their genomes, the German scientists were able to show that homo sapiens inter-bred with the Denisovans while on their way to colonize south Asia and Australia. "When homo sapiens expanded out of Africa around 100,000 years ago or less, obviously the east Asians incorporated a little gene flow from these Denisovans," Professor Groves said.

While the scientists who conducted the study had no Aboriginal Australian samples, Professor Groves said Melanesians, Papuans and Aboriginal Australians formed a group of related peoples. "It does seem likely that if there were Aboriginal samples, they too would prove to have some gene flow from the Denisovans," he said. Professor Groves suggested the remains unearthed at Denisova might be analogous to a skull discovered by farmers in China's Guangdong province in 1958.

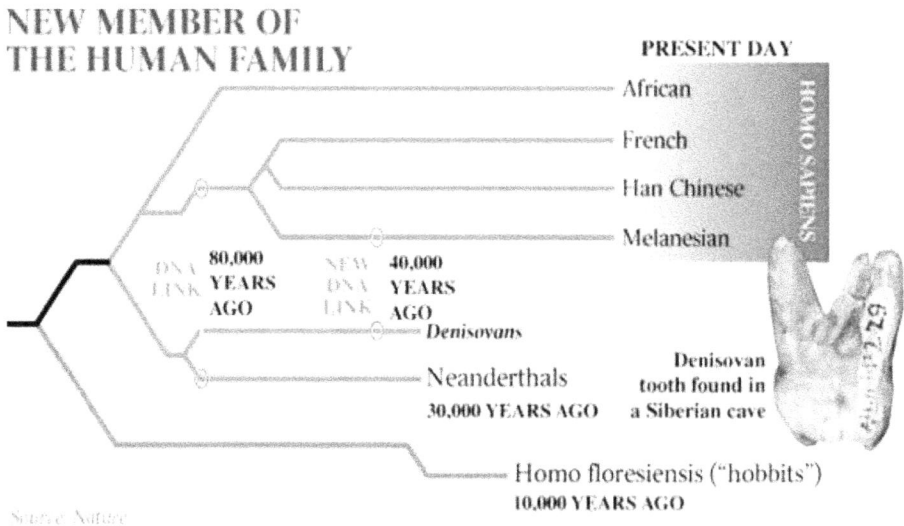

NEW MEMBER OF
THE HUMAN FAMILY

PRESENT DAY

African

French

Han Chinese

Melanesian

HOMO SAPIENS

80,000 YEARS AGO
DNA LINK

NEW 40,000 YEARS AGO
DNA LINK

Denisovans

Denisovan tooth found in a Siberian cave

Neanderthals
30,000 YEARS AGO

Homo floresiensis ("hobbits")
10,000 YEARS AGO

Source: Nature

Figure 11-6 : Denisovan, Homo Sapien
& Neanderthal Family Tree

The director of the Australian Centre for Ancient DNA at the University of Adelaide, Alan Cooper, said gene-sequencing technology was revolutionizing the study of human evolution. "We've moved from testing existing hypotheses that have been based on traditional science -- fossils and archeology, where we're just testing things that were already known -- to actually making new hypotheses ourselves," Professor Cooper said. "This thing wasn't even supposed to exist -- there were no remains for it, no knowledge of it -- and yet the genetics is now demonstrating complete chunks of the human tree that were not even known."

Professor Cooper said this, and other discoveries showed Asia had a "whole bunch of human history going on that we know nothing about". Asked whether gene-sequencing could redraw the tree of human evolution in the next five years, he replied: "I wouldn't be totally surprised."

CHAPTER 12
Homo Sapiens (Modern Humans)
12.1 Modern Man & Conventional Wisdom

For much of the last century, the conventional wisdom about the origin of modern humans was that the transformation from archaic Homo populations to modern humans took place more or less independently in each of the main regions of the Old World, that is Africa, Europe, and Asia. So, for example, in Europe, the Neanderthals would have evolved into European modern humans, and in Asia late surviving H. erectus would have evolved into Asian modern humans. In its extreme form, this multiregional hypothesis embraced the now thankfully discredited notion that geographical variants of modern humans (the term race' has little, or no, biological meaning with respect to modern humans) were separate species with distinctly different evolutionary histories. A weaker form of the multiregional hypothesis was espoused by researchers such as Franz Weidenreich (who had played a critical role in the analysis of the remains of Homo erectus from Zhoukoudian). This combined the hypothesis that regional variants of archaic Homo had each evolved into modern humans, with the proposal that subsequent to their independent evolution the differences between these regional variants were eventually reduced by gene flow (either by migration or by inbreeding) between the regions. Nonetheless, contemporary supporters of this weak multiregional hypothesis (WMRH) argue that despite gene flow each region has kept enough of its own character to make regional populations of modern humans distinctive and recognizable. They support the WMRH because they see morphological evidence of continuity between pre-modern Homo and modern human populations in each of the major regions of the world. For example, they claim dental and cranial evidence links H. erectus and modern Australians, and that a distinctive facial morphology links the Neanderthals and modern Europeans.

In this scenario for the evolution of modern humans it would be difficult to draw a line between, say, Neanderthals and early modern humans in Europe, and between H. erectus and early modern humans in Asia. Supporters of the WMRH argue that these gradations, together with the melding effect of the gene flow that has occurred between geographical regions, justify including H. erectus and all the regional hominin variants that came after it in a single species. If there were to be a single species for H. erectus and all subsequent hominins, then that species would have to be Homo sapiens. Linnaeus' species name for modern humans has historical priority over all the other names (e.g., H. neanderthalensis and H. heidelbergensis) subsequently given to pre-modern Homo species. Eurocentrism in palaeoanthropology The first discovery of a fossil modern human to be published was probably the recovery of the skeleton of the Red Lady' (the bones were stained with red ochre) from a cave at Paviland on the Gower Peninsula, just west of Swansea, Wales, in 1822–3. However, the discovery that is nearly always cited as the first fossil evidence of modern Homo (i.e., Homo sapiens) in Europe was made in 1868 at the Cro-Magnon rock shelter at Les Eyzies in the Dordogne, France. The apparent historical priority of Cro-Magnon, combined with the archaeological evidence of sophisticated small stone awls, and needles and fishhooks made from bone recovered from European sites, suggested to many researchers that continental Europe was not only the cradle of modern civilization, but that it was also the birthplace of our own genus, Homo, and our own species. The strong' and weak' versions of the multiregional and recent out of Africa models for the origin of modern Homo A challenge to Eurocentrism The preconception that Europe was the place where modern humans evolved was challenged by two developments. The first was the recognition, beginning in the latter part of the 19th century, and intensifying in the second quarter of the 20th century, that there was fossil evidence of human ancestors more primitive than Neanderthals in Asia. Subsequently, of course, came the realization that the early phase of hominin evolution most likely occurred in Africa. The second development took place in the University of Cambridge in England. It started in the 1930s with the discovery by Dorothy Garrod, a distinguished Cambridge archaeologist, of fossil remains resembling modern humans in caves on

Mount Carmel in what was then Palestine. The Mount Carmel discoveries, together with the recovery of modern human-like fossils and evidently ancient stone tools in Kenya by Louis and Mary Leakey, and in Egypt by Gertrude Caton Thompson (both also affiliated with the archaeology department at the University of Cambridge) began to convince the more outward looking European archaeologists that important events in both the early and the later stages of human evolution may have taken place outside of Europe. In 1946 Dorothy Garrod introduced a course called World Prehistory' into the undergraduate archaeology course at Cambridge and her successor, Grahame Clark, continued in the same vein by encouraging his graduate students to excavate in Africa. The point of this diversion into prehistory is to make the point that by the 1950s and 1960s some students of human evolution were already comfortable with the idea that important events in the evolutionary history of modern humans may have taken place outside Europe. Discoveries, new dates, and molecular evidence In the 1980s three lines of evidence combined to prompt some researchers to contemplate the radical proposition that Africa, far from being an evolutionary sideshow and a cultural backwater, may have been the birthplace of modern humans and of modern human behavior. The first of the three new lines of evidence was the redating of the collections of hominin fossils in the Levant. This made it clear that instead of the Neanderthal fossils from Kebara and Amud predating the more modern human-looking fossils from Skuhl and Qafzeh, it was the other way round. The modern-looking fossils from Qafzeh were older than the fossils from Kebara and Amud that evidently belonged to an archaic Homo species. This meant that researchers could not use dating evidence to make the case that Neanderthals evolved into modern humans. The second line of evidence was the discovery of modern human-looking fossils in southern Africa and in Ethiopia. The most influential discovery was made in 1968 at Klasies River Mouth in South Africa. Here researchers had uncovered skull fragments that looked for all the world as if they might have belonged to a modern human, yet they were perhaps 120 KY old. A similar date was also initially suggested for a modern human-looking cranium from a locality called Kibish in the Omo Region in southern Ethiopia.

On rather weak biochronological evidence the Omo I cranium had been dated to c.120 KYA, but a recent attempt to date the Omo I cranium using isotope dating has suggested a substantially older date, closer to 200 KYA. A collection of fossils from Herto, another Ethiopian site, also suggests that modern human-like fossil hominins were present in Africa between 200 and 150 KYA. The third line of evidence came not from palaeoanthropology, but from the application of molecular biological methods to the study of modern human variation. The pioneering study applying these methods was published in 1987 by Rebecca Cann, Mark Stoneking, and Allan Wilson, molecular biologists at the University of California at Berkeley. For several reasons it focused on mtDNA and not on nuclear DNA. Mutations occur in mtDNA at a faster rate than they do in nuclear DNA, and unlike nuclear DNA mtDNA does not get reshuffled between chromosomes when germ cells divide. Nor does it have all of the innate mechanisms for DNA repair that are found in the nucleus. This may contribute to its higher mutation rate, and account for the observation that once mutations occur in mtDNA they tend to persist. The Cann et al. study compared mtDNA from 147 modern humans, 46 from Europe, North Africa, and the Near East, 20 from sub-Saharan Africa, 34 from Asia, 26 from New Guinea, and 21 Australians. The researchers found 133 different versions of mtDNA. They arranged them in the shortest tree that connected all the variants while minimizing the number of mutations. The shape of the tree they constructed from their results was striking, as was the geographical distribution of the differences between the various types of mtDNA. The tree had a deep African branch and a second branch that contained the mtDNA variants found in people from outside sub-Saharan Africa. The variation in mtDNA was not even across the tree. There was more variation within the sub-Saharan African branch of the tree than in the rest of the world put together. Not only that, most of the mtDNA variants seemed to have had an African origin or Mitochondrial Eve. These results could mean one, or both of two things.

First, modern humans had been in Africa longer than anywhere else in the world. Second, that the population size of modern humans in Africa was larger than that in the rest of the world combined. This makes sense, for the more people there are, the more likely it is that mutations will occur. Cann and her colleagues made three other claims in their paper. First, because it was then widely assumed that mtDNA differences were not under the influence of natural selection (i.e. the mutations are _neutral') and because most mtDNA differences do not affect the function of the cellular machinery genes they code for, this means that any differences in mtDNA that have accumulated between two population samples are simply a function of how long those two populations have been undergoing independent evolution. Second, Cann et al. suggested that the differences between the sub-Saharan and the non-sub-Saharan populations of modern humans would have taken about 200 KY to accumulate, and therefore their prediction was that modern humans originated in Africa around 200 KYA. Third, they claimed that the distribution of the mtDNA variants suggested that when modern humans left Africa, they did not interbreed with any of the archaic populations they must have encountered as they moved into the other main regions of the Old World. Cann and her colleagues claimed that only African archaic Homo populations contributed to the gene pool of modern humans, and thus also they supported the corollary, which is that archaic hominins in other parts of the world made no contribution to the modern human genome. In effect, Cann and her colleagues claimed that all post-200 KY-old hominins only have African genes. Because you inherit the vast majority of your mtDNA from your mother, the evolutionary history of mtDNA is effectively a history of maternal inheritance. Thus, it is not surprising that either the press, or the researchers, came to call Cann et al.'s interpretation the Mitochondrial Eve' hypothesis. It was called that because one of its implications is that the mother of all humanity was a c.200 KY-old African female. I will refer to it as the strong recent out of Africa (SROAH) hypothesis, but as we will see below most researchers who support a recent out of Africa' model for modern human origins now support a less extreme version.

So, the battles lines were drawn. In the red corner' the weak multiregional hypothesis (WMRH), and in the blue corner' the weak recent out of Africa hypothesis (WROAH). Remember that some researchers who were unwilling to support the strong version of the multiregional hypothesis were more inclined to support a weaker interpretation that included gene flow between regions. Similarly, when other researchers tried to repeat Cann et al.'s results using more up-to-date molecular methods and more rigorous statistical techniques, they came up with different results. These still pointed to Africa as the origin of a substantial amount of modern human mtDNA variation, but several of these studies suggested there was evidence that pre-modern Homo from outside Africa also contributed to the modern human mtDNA genome. The male and the nuclear perspectives While researchers were working on ways to refine the evidence for modern human origins that could be extracted from regional variations in modern human mtDNA, other research groups had set about tackling other parts of the genome. One of the parts of the nuclear genome they paid particular attention to is the DNA from the part of the male, or Y, chromosome, which has no equivalent on the female, or X, chromosome. Because it has no female counterpart, the DNA on that part of the Y chromosome does not get reshuffled during germ cell division: the technical term for it and the mtDNA is that they are both non-recombining' regions of the genome. So, this part of the Y chromosome DNA is like mtDNA except that it is transmitted from one generation to the next by males and not by females. The results from studies of the Y chromosome were like those from the mtDNA studies. Twenty-one out of twenty-seven Y chromosome variants originated in Africa, and there was more variation in the Y chromosome of Africans than in all the people from other parts of the world, thus the mtDNA results were no flash in the pan'. Much the same results have come from studies of nuclear genes, but like those in mtDNA and in the Y chromosome, nuclear gene studies are providing evidence of admixture between archaic and modern human genotypes. The predominant message from DNA studies, be it from mtDNA, the Y chromosome, or the regular autosomal nuclear genome, is that most, but certainly not all, modern human genes originated in Africa.

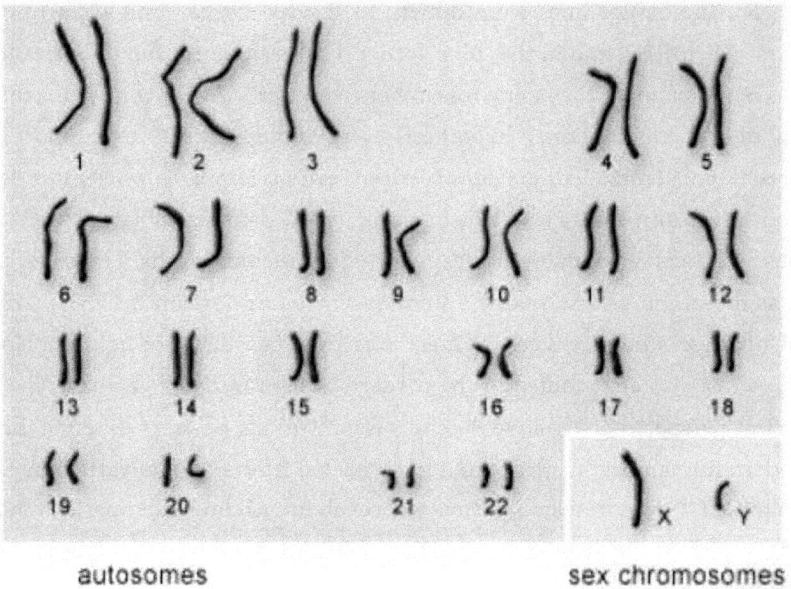

autosomes   sex chromosomes

U.S. National Library of Medicine

Figure 12-1 : Human Chromosomes comprising the DNA

Another is that for the past 2 MY Africa seems to have been the source of pulses' of hominin evolutionary novelty. The first pulse was the emigration of a H. ergaster-like hominin, then a H. heidelbergensis-like hominin, and then perhaps several waves of migration of modern human-like hominins, perhaps not looking very different, but with different cultural capacities and skills. It is now generally agreed that modern humans are derived from a relatively recent, c.50–45 KYA migration out of East Africa. One researcher, Alan Templeton, whose important contribution pointed out the evidence for a series of migrations, gave his paper the apt title Out of Africa Again and Again'.

12.2 Migration or Gene Flow?

Novel genes can reach beyond Africa in two ways. People can take them with them when they migrate, or they can transmit them by interbreeding. The latter mechanism involves Africans interbreeding with people in an adjacent region of the Old World, these people then in turn interbreeding with other people further away from Africa, and so on. The genes are transmitted rather like the baton in a relay race. This is the type of gene transmission implied in one of the more recent theories about modern human origins. It is called the diffusion wave hypothesis', and it suggests that novel genes spread in waves. It is consistent with the results of a recent study that shows a strong correlation between genetic distance' and the actual distance in miles of the shortest overland route between where the sample of modern humans was from and the African continent. Modern humans beyond Africa There are two discussions about the arrival of modern humans anywhere beyond Africa, be it in Europe, or anywhere else. One concerns the arrival of modern human-looking people themselves, in other words the earliest fossil evidence of modern humans. The other discussion concerns the arrival of modern human behavior, in other words the earliest archaeological evidence of people doing things that archaeologists are satisfied that only modern humans would have been able to do. Not surprisingly, the discussions about what constitutes modern human behavior are more spirited than those surrounding what constitutes modern human morphology.

Once paleoanthropologists managed to escape from the trap of equating modern human morphology with the morphology of modern Europeans, it became easier for them to recognize modern humans in different parts of the world. Archaeologists have also recognized that there is more to modern human behavior than what our ancestors were doing in Europe starting c.40 KYA. For example, the alleged lack of cave art in Africa was sufficient to dismiss Africa as a potential source of modern human behavior. There are two good reasons to reject this argument. First, there is cave art in Africa; archaeologists had not been looking hard enough. Second, to have cave art you need caves, and in many parts of Africa there are no caves. Modern humans in Europe The earliest fossil evidence of modern humans in Europe comes from a site in south-east Europe called Pestera cu Oase in Romania.

The main morphological and behavioral differences between modern humans and Neanderthals which is dated to around 35 KYA, and we know that modern human-looking people had reached England, at Kent's Cavern, by about 30 KYA. The earliest evidence of modern human behaviour in Europe currently comes from sites in Bulgaria called Bacho Kiro and Temnata, dated to between 43 and 40 KYA, and by just less than 40 KYA there are many sites across Western Europe that show evidence of modern human behaviour. Modern humans in Europe overlapped with the Neanderthals for around 10 KY or less, depending on the location. The most recent evidence for Neanderthals comes from sites such as St Césaire in France, Zaffaraya in Spain, and Vindija in Croatia that are all dated to c.30 KYA. Modern humans in Asia: Sahul and Oceania Researchers have suggested that modern humans may have occupied one, or more, parts of Sahul, the landmass that includes Papua New Guinea, Australia, and Tasmania, by 40 KYA.

With so much water locked up in polar ice caps and glaciers, land that is part of the continental shelf and which is now submerged would have provided dry connections between landmasses that are today separated by water. If hominins were in Sahul by 40 KYA, then they must have been in Sunda, the landmass that includes mainland South-East Asia and the present-day islands that make up Indonesia, sometime before that. If the late dates for the last H. erectus fossils in this region, from Ngandong, Java, are correct, then there would have been overlap between modern humans and late H. erectus. But the discovery of Homo floresiensis, a dwarfed' form of Homo erectus that persisted until 18,000 years ago on the island of Flores is a reminder that temporal overlap does not necessarily mean that their ranges overlapped. Different kinds of hominins could have lived on separate islands and not necessarily have come into contact with one another. These early modern humans in Sunda must have been able to travel on rafts, or some other form of craft, and to have managed well enough to spend at least several days at sea in order to cross the open water between Sunda and Sahul. By 35–30 KYA, modern humans in the Pacific region were skilled enough as seafarers to reach many remote islands in Oceania including Timor, the Moluccas, New Britain, and New Ireland.

Modern humans in Sahul The existing hominin fossil record suggests that modern humans were the only hominins to enter the region we call Sahul, so there is no question of overlap with earlier groups. The time of the initial arrival of modern humans in Australia is unknown. Fossil evidence indicates that they might have arrived by 50 KYA, but they were certainly there between 40 and 35 KYA when the climate was wetter than it is today. Modern human fossils in Australia show substantial morphological variation. The people living at sites around Lake Mungo had steep foreheads, taller brain cases, and flat faces, while people at Kow Swamp and Coobool Creek in Northern Victoria had more sloping foreheads, lower brain cases, and projecting faces. Some researchers interpret these morphological differences as evidence of more than one wave of immigrants, but others see no more variation than one would expect if a new species dispersed across a large new territory such as Australasia. Modern humans in the New World There were three routes from the Old World to the New, across the Bering Straits, island hopping from one Aleutian island to another, or across the Atlantic. Today all three require a sea voyage, but for several periods during the past 40–30 KY the fall in sea level and the thick ice caused by the intensely cold conditions would have closed the Bering Straits, linked some of the Aleutian Islands and would have made even a transatlantic voyage less formidable. The problem in all three cases was the intense cold those making the journey would have experienced.

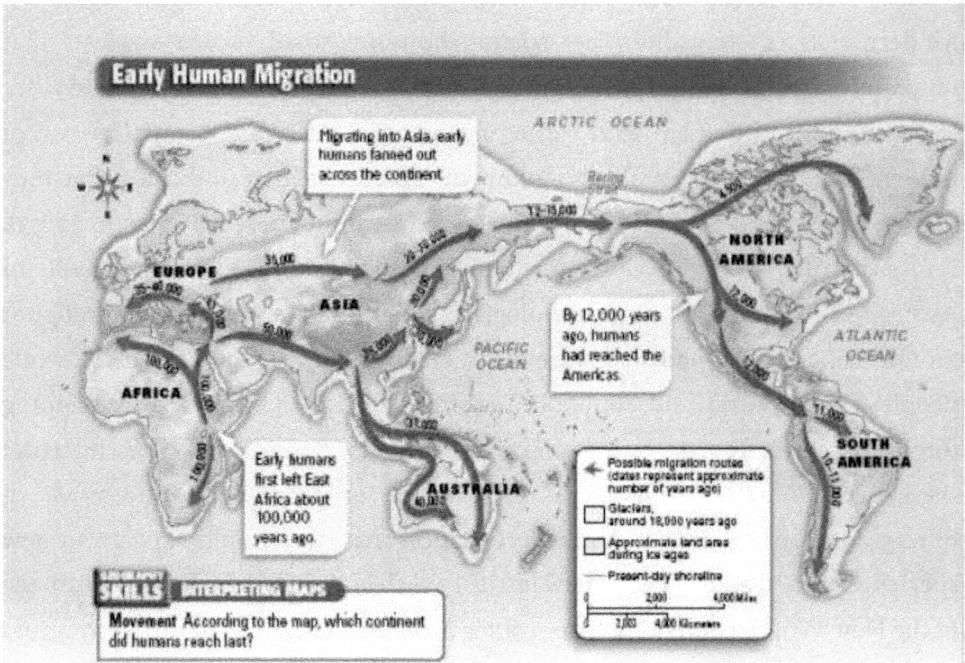

Early Human Migration

Migrating into Asia, early humans fanned out across the continent.

By 12,000 years ago, humans had reached the Americas.

Early humans first left East Africa about 100,000 years ago.

Possible migration routes (dates represent approximate number of years ago)
Glaciers, around 18,000 years ago
Approximate land area during ice ages
Present-day shoreline

SKILLS INTERPRETING MAPS
Movement According to the map, which continent did humans reach last?

The first evidence for modern human occupation within the Arctic Circle is 27 KYA, and by 15 KYA there is evidence of long-term occupation. During this period it is possible that modern humans following migrating herds of mammoths ventured unwittingly into the New World, but we do not find any evidence of a modern human occupation site in Alaska until 12 KYA. The conventional wisdom is that the immigrants made their way south along a relatively ice-free corridor in Alaska and western Canada, and then went on to populate all of North, Central, and South America relatively rapidly. However, there is remarkably little evidence of human occupation along what is presumed to be the route south. And some New World archaeologists use this negative evidence in support of other scenarios, including one suggesting that the first occupants of the New World may have travelled there directly from Europe. The best-known archaeological evidence for modern humans in the New World is the Clovis culture, characterized by distinctive stone tools called Clovis points.

The oldest Clovis sites are dated to slightly before 11 KYA, and not long after this there is abundant evidence of Clovis points over most of the unglaciated regions of North America. For a long time, archaeologists accepted the Clovis sites as the earliest evidence of modern humans in the New World. But more recently researchers have claimed they have unearthed evidence of a stone industry that is more primitive than the Clovis. The best known of these preClovis sites in North America are Duktai in Alaska, Meadowcroft in Pennsylvania, Cactus Hill in Virginia, and Topper in South Carolina. In South America the best-known sites are Taima-Taima in Venezuela, Pedra Furada in Brazil, and Monte Verde in Chile. Most of these sites are dated using relative methods, but the dates of two sites, Meadowcroft and Monte Verde, are reasonably reliable. Meadowcroft's radiocarbon dates indicate it was inhabited by at least 14 KYA, and perhaps as early as 20 KYA. Monte Verde provides excellently preserved evidence of modern human behaviour in South America around 12.5 KYA. There is even preservation of the cords used to tie hides to poles, and the remains of a dwelling that was big enough to have housed 20–30 people.

Monte Verde was occupied year-round; thus, it is the earliest evidence of a semi-permanent occupation site in the New World. A persistent problem with the hypothesis that the Clovis people were the first to occupy the New World is that most of the Clovis sites are in the eastern part of the United States and Canada. If the Clovis people came across what was then the Bering land bridge how can one explain the distribution of the sites? An archaeologist, Dennis Stanford of the Smithsonian Institution's National Museum of Natural History, has proposed a radically different hypothesis. This suggests that the first inhabitants of the New World were modern human groups from Spain. The author points out that similarities between the Iberian Solutrean tradition and some of the flakes in the Clovis toolkit support an Iberian' rather than a Siberian' source for the modern human settlement of North America. It is likely there were several migrant streams of modern humans into the New World. Different groups, over different periods, arrived and settled, and each made their own contribution to the genetic and cultural diversity of New World populations.

No matter when, where, and how modern humans arrived in the New World, it did not take them long to spread rapidly over a diverse range of environments. The recent announcement of the discovery of 40 KY-old human footprints in Mexico has added yet another contentious claim to an already contentious topic. Points to watch. Researchers will be keen to find more sites in Africa that date to between 300 KYA and the present, and to find ways of dating them reliably. Some researchers are confident that H. erectus evolved into H. sapiens via populations with crania like those from Kabwe in Zambia and Bodo in Ethiopia. But this may be an over-simplistic interpretation. Researchers also need to keep looking in the regions immediately adjacent to Africa for hominin evidence. As the technology for gene sequencing continues to improve, more genes will be sampled, and larger numbers of individuals will be sampled from each region. Researchers will be focusing on nuclear genes to see if non-African pre-modern Homo genes made a very minor, or a more significant, contribution to the modern human gene pool. Researchers interested in the later stages of human evolution are still unsure about the connections between morphology and behavior. Were changes in cranial shape associated with cultural changes?

For example, at what stage did modern Homo begin to use complex spoken language, and could we tell they had reached that stage just by looking at the shape and size of the brain? Was the shift to making small, complex, stone tools the result of changes in the hands, or were these innovations entirely cognitive?

CHAPTER 13
MIGRATIONS OF MODERN MAN and DIVERSITY
13.1 The Migrations of Modern Man and Racial Diversity

The fossil record places human origin in Africa, but science continues to search for details about the incredible journey that took Homo sapiens to the far reaches of the Earth. How did each of us end up where we are? Why do we have such a wide variety of colors and features? Such questions are even more remarkable in light of genetic evidence that we are all descended from a common African female ancestor (eve) who lived only 140,000 years ago.

Through the eons of time, the full story remains clearly written in our genes. When DNA is passed from one generation to the next, most of it is recombined by the processes that give each of us our individuality. But some parts of the DNA chain remain largely intact through the generations, altered only occasionally by mutations, which become —genetic markers. These markers allow geneticists to trace our common evolutionary timeline back many generations.

Different populations carry distinct genetic markers. Following the markers through the generations reveals a genetic tree on which today's many diverse branches can be followed backwards to their common African root.

The markers in our genes allow us to chart the ancient human migrations from Africa across the continents. Through these markers, we can see living evidence of an ancient trek to populate the globe.

Australia, South Asia, and China

Eventually when ancient Homo-sapiens-sapiens (Modern Man), began to leave Africa: There were two great migrations East. The first of these occurred about 60,000 B.C. This group followed a coastal route across Southern Arabia and Southern Asia, then "Island Hopped" to Australia across the frozen glacier ice at the end of the Great Ice Age. (Hainesworth) The second wave of migration occurred about 50,000 B.C. These may have been big game hunters who after crossing Southern Arabia, followed an inland route in search of game; they reached China by about 45,000 B.C.

In each new land that they found, some stayed behind and made it their home. In India they created the Indus Valley Civilization, in Burma and Thailand they eventually created the Ban Chiang and Mon civilizations. In Cambodia, they created the Khmer kingdoms of Funan and Angkor. In Vietnam, they created the Champa civilization. In Indonesia, they created civilizations in Malaya and the Indonesian archipelago. In Japan, they created the Jomon and Ainu cultures. In China, they were the creators of the first civilizations, the Xia and Shang civilizations.

Naro bushman, Central Kalahari , Botswana

Figure 13-1 : A Bushman family in the Kalahari desert

A note of interest: The Bushman of the Kalahari desert in South Namibia. Aka: The San People. Have by genetic analysis been determined to be the closest to the original Homo-sapiens sapiens in genetic makeup, and thus, one of the world's Oldest Humans!

Also Note : The Complete Khoisan and Bantu genomes from southern Africa have just been completed and published (Received 11 August 2009; Accepted 6 January 2010)

Figure 13-2 : Depiction of early Homo Sapien leaving Africa

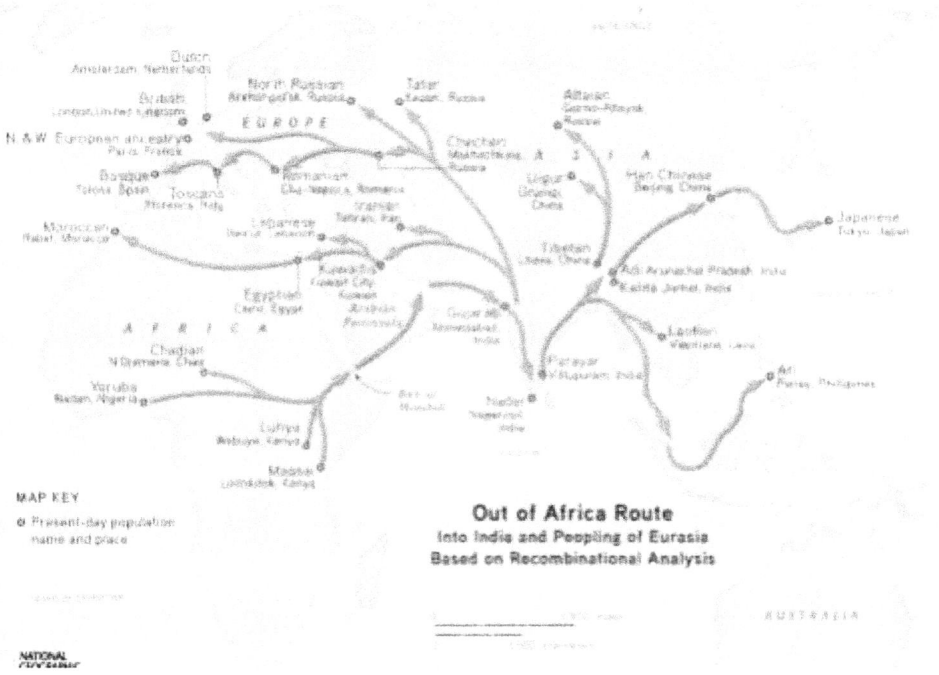

Figure 13-3 : Map of Out of Africa Migration of Modern Humans

13.2 Environmental Adaptation & Racial Diversity

Europe

Over time, the last obstacle to Modern Mans Migrations started to disappear: The Great Glacial Ice Sheets that once covered most of Europe, started to melt. The first Modern Human to enter Europe, at about 45,000 B.C. was the Khoisan type African, commonly called "Grimaldi Man". One known entry point into Europe for Grimaldi Man, was the straits of Gibraltar, which were passable because of the lowered Sea level caused by the "Ice Age". Once in Europe, Grimaldi Man continued his migrations, and came to eventually inhabit all of Europe and Northern Asia. The Easternmost limit of his range appears to have been the settlement known as Mal'ta in Siberia Russia, just north of Mongolia.

Though Grimaldi Man is known to have established settlements as far south as Catal Huyuk in Anatolia. There is uncertainty as to whether it was Grimaldi descendants, or a different group, such as those who settled North Africa and the Middle East, that can be credited with the creation of the original Southern European civilizations, especially those in Italy and Greece. Likewise, there is uncertainty as to what part, if any, Grimaldi people played in the creation of the founding Xia and Shang civilizations of China. However, by virtue of their settlements at Mal'ta Siberia, it is certain that they were neighbors of the Africans who created those civilizations.

13.3 North Africa and the Middle East

Thought the migrations of Modern Man into Europe, Asia, and the Americas, (and the civilizations created in those places), is covered here. The bulk of our presentation deals with Man's migration into North Africa and the Middle East where he creates great civilizations in Nubia, Egypt, Canaan, Mesopotamia, Iran, and India. Links at the bottom of the page, guide you through the presentation.

13.4 The P-Gene Hypotheses

Although this is a sensitive subject for some people, it will be addressed here in order for the reader to have a clear understanding of genetic phenotypic diversity with respect to the outward appearances of modern humans. (DPH)

Accepting the hypotheses of the San as the first Modern Humans also allows for easy explanation of African skin colors. The UV index of South Africa reaches the maximum of eleven in only four to six months of the year. Therefore "Brown Skin" provides ample protection from the Sun. But to the north at the Equator, the UV index is a constant maximum of eleven all year long. That leaves just one more Human attribute to be explained : Straight Hair. Previously we have seen that Albinism (a defective "P" gene) actually straightens the hair. And we have also seen that the effects of a defective "P" gene are not always universal. Thus a person with a defective "P" gene might only have it effect the eyes or only the hair.

Figure 13-3 : Ethiopians of diverse ethnicities

Nigerian parents in London with their third Child

Figure 13-4 :African Nigerian family with their newborn

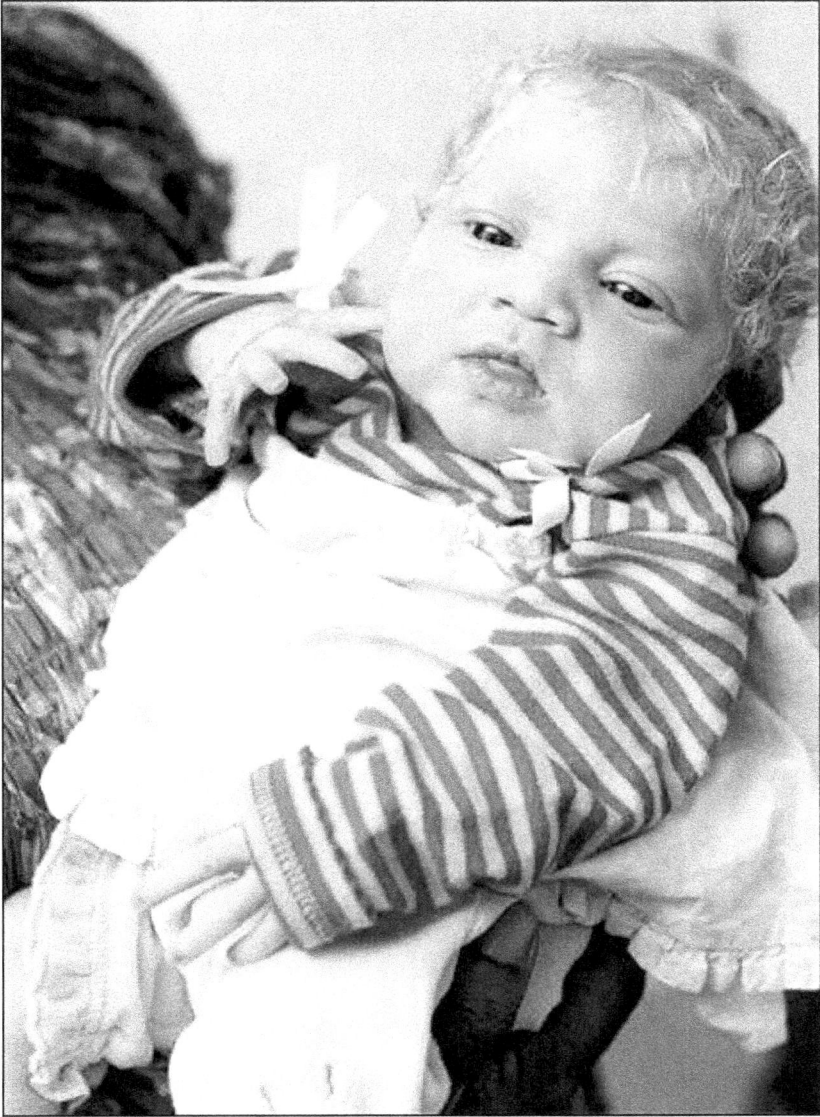

Figure 13-5 :African Nigerian family's newborn

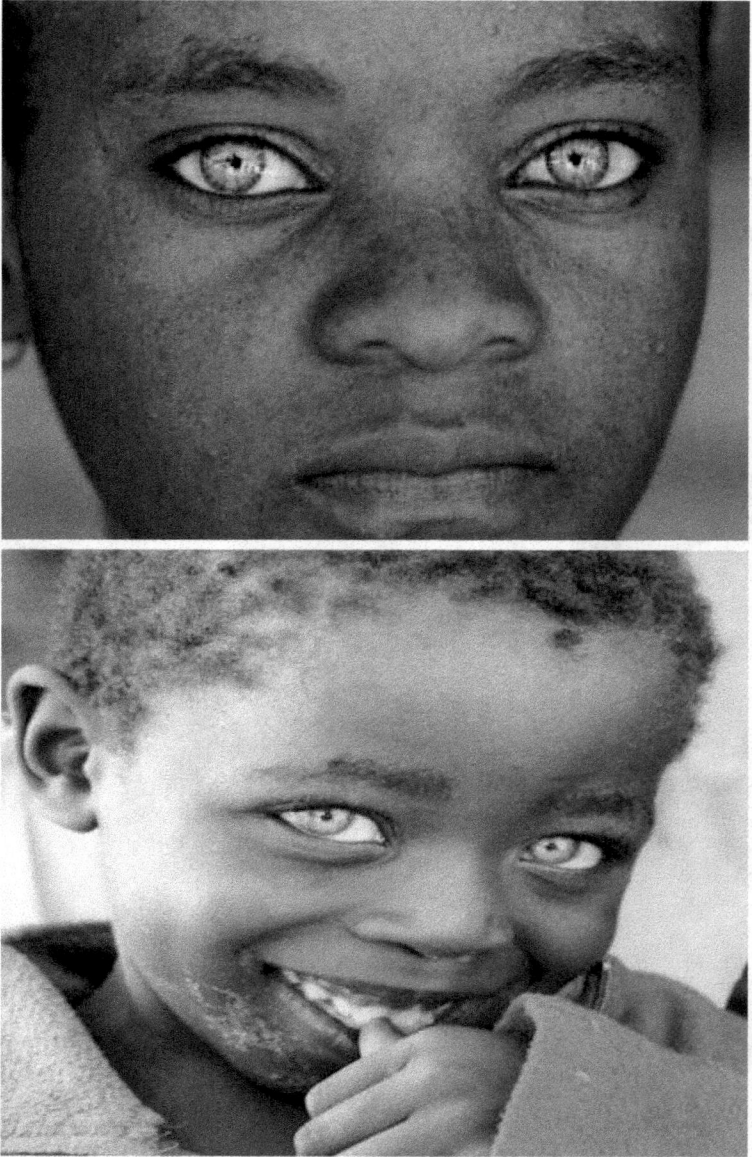

Figure 13-6 :Children of Black African heritage with light eyes

And of course, when the effects of a defective "P" gene are universal, it gives rise to the Dravidian Albino: formerly of Central Asia, and now of Europe.

13.8 Some Common Ancestry

Genetic data shows that the biochemical systems of Asian and European populations, appear to be more similar to each other, than they are to African populations. Thus, Asians (Mongols) and Europeans (Caucasians) may have shared a common ancestry with each other, some 40,000 years ago and a common ancestry with African populations before that. The Out of Africa (OOA) migration, which took Africans into Asia, occurred at about 50,000 B.C. The modern Mongol shows great affiliation with San Africans in body type and facial features, thus the presumed genesis below (see figure 13-14).

Figure 13-7 : A Solomon Islander child with light colored hair

Note : These mutated genes effecting the eyes, hair texture and color pigmentation must have been better suited for colder climates such as Europe and parts of Asia. (DPH)

294

Scientists have finally figured out how some of the dark-skinned inhabitants of the Solomon Islands have naturally blonde hair. Researchers used to believe the blonde hair came from interaction with European people, however, a group from Stanford has detected a genetic difference in the blondes. They swabbed the cheeks of 85 people, 43 with blonde hair, to compare their DNA to that of people with darker hair and found a chromosomal difference caused the blonde hair. The researcher identified a change in the gene TYRP1, which affects pigment in humans and mice, as the cause.

The scientists consider the effect to be very unique. —The mutation is at a frequency of 26 percent in the Solomon Islands, is absent outside of Oceania, represents a strong common genetic effect on a complex human phenotype, and highlights the importance of examining genetic associations worldwide,‖ said the abstract of the report. The team was stunned by their findings. —They have this very dark skin and bright blond hair. It was mind-blowing, Sean Myles, one of the researchers, told the Daily Mail. —As a geneticist on the beach watching the kids playing, you count up the frequency of kids with blond hair, and say, Wow, it's 5 to 10 percent'.‖ Eimear Kenny, co-author of the study, has similar feelings. —Within a week we had our initial result. It was such a striking signal pointing to a single gene — a result you could hang your hat on, he said.

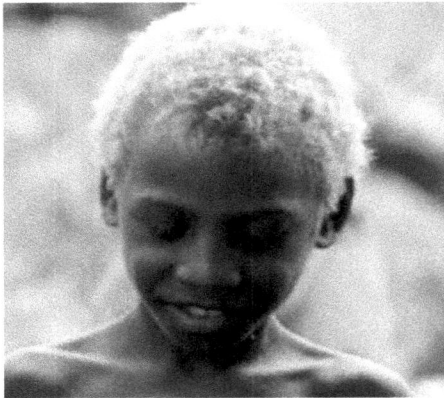

Figure 13-8 :A Solomon Islander Boy

Figure 13-9 : A crowd of Solomon Islander Men and Boys

Figure 13-10 : Solomon Islander Boys

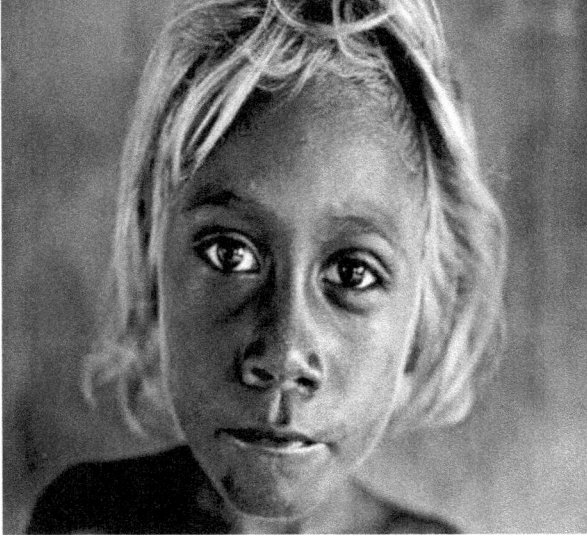

Figure 13-11 :A Micronesia Islander Boy

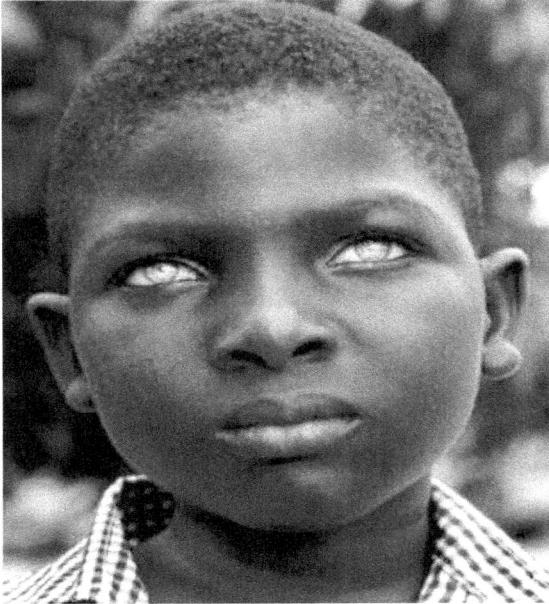

Figure 13-12 : An African Boy

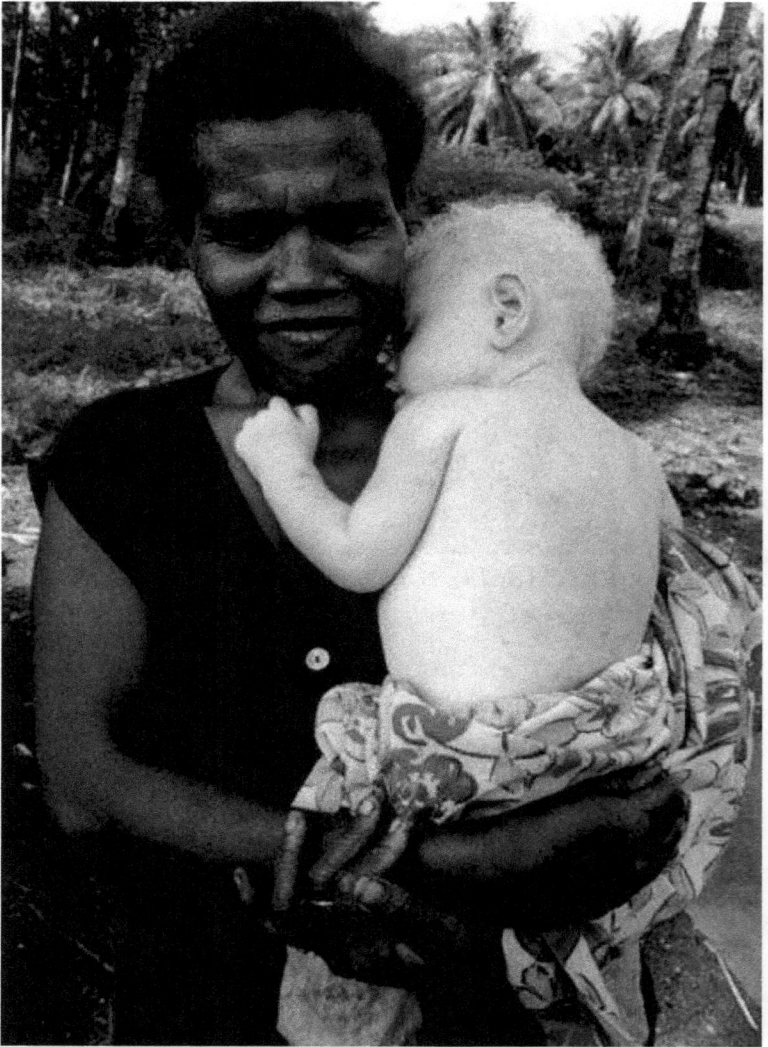

Figure 13-13 : A mother With Her Child

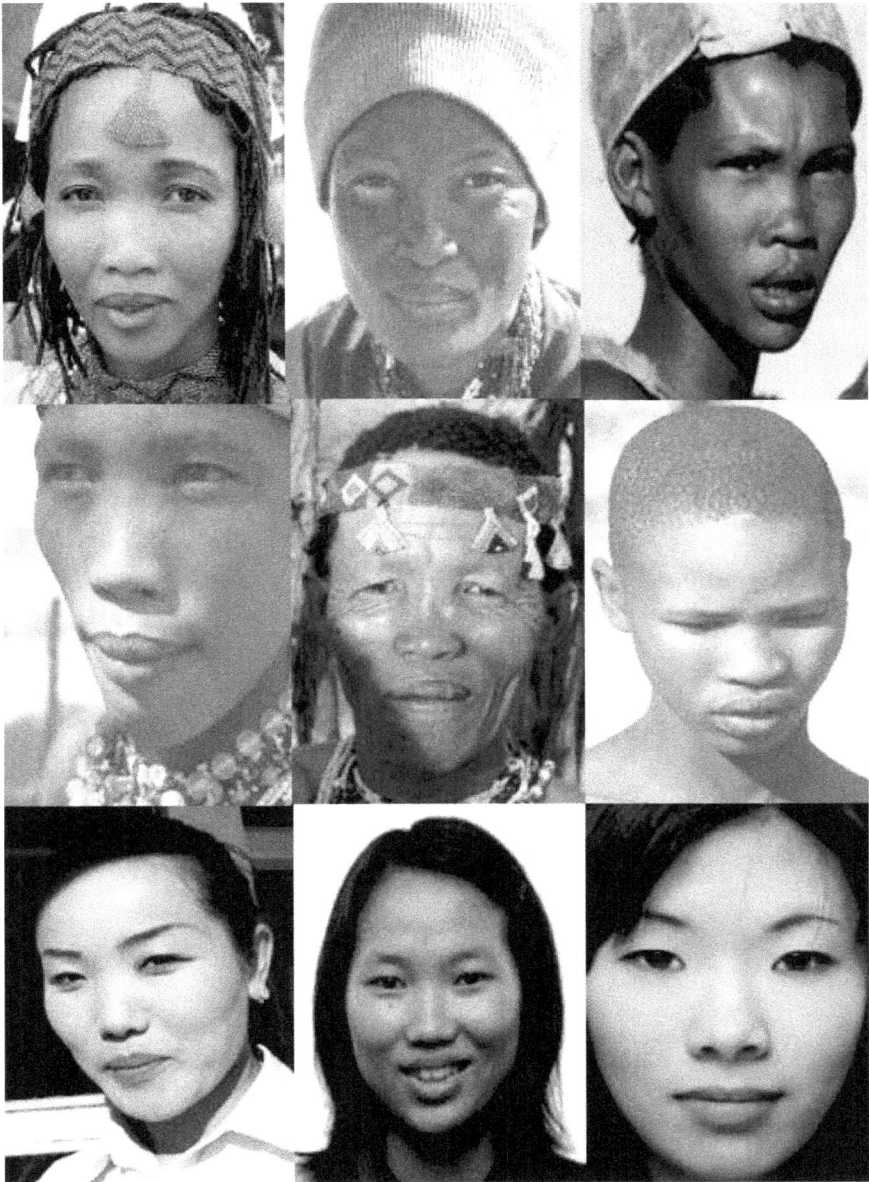

Figure 13-14 : Khoisan Africans & Asians with like Facial Features

299

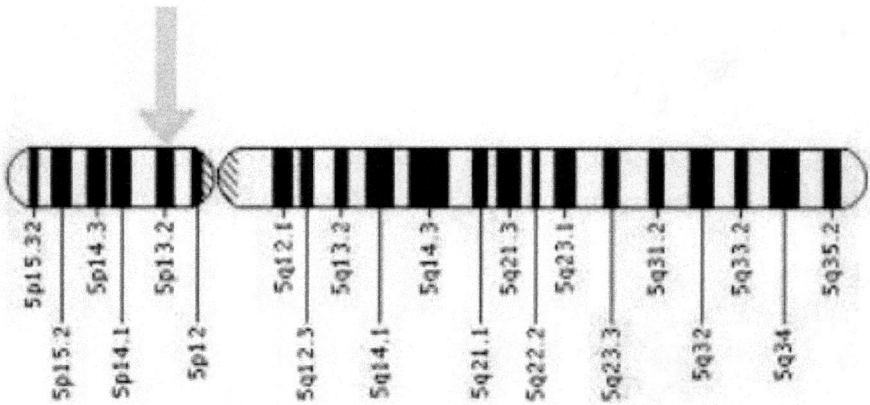

Figure 13-15 : A single gene called the —P-gene‖ as depicted in the above figure has been identified to be the main cause for hair color and skin tone alteration.

13.9 Albinism Albinism is a defect of melanin production that results in little or no color (pigment) in the skin, hair, and eyes.

Causes

Albinism occurs when one of several genetic defects makes the body unable to produce or distribute melanin, a natural substance that gives color to your hair, skin, and iris of the eye.

The defects may be passed down through families.

There are two main types of albinism:

- Type 1 albinism is caused by defects that affect production of the pigment, melanin.
- Type 2 albinism is due to a defect in the "P" gene. People with this type have slight coloring at birth.

The most severe form of albinism is called oculocutaneous albinism. People with this type of albinism have white or pink hair, skin, and iris color, as well as vision problems.

Another type of albism, called ocular albinism type 1 (OA1), affects only the eyes. The person's skin and eye colors are usually in the normal range; however, an eye exam will show that there is no coloring in the back of the eye (retina).

Hermansky-Pudlak syndrome (HPS) is a form of albinism caused by a single gene. It can occur with a bleeding disorder, as well as with lung and bowel diseases.

Other complex diseases may lead to loss of coloring in only a certain area (localized albinism). These conditions include:

- Chediak-Higashi syndrome (lack of coloring all over the skin, but not complete)
- Tuberous sclerosis (small areas without skin coloring)
- Waardenburg syndrome (often a lock of hair that grows on the forehead, or no coloring in one or both irises)

SYMPTOMS

A person with albinism will have one of the following symptoms:

Absence of color in the hair, skin, or iris of the eye
Lighter than normal skin and hair
Patchy, missing skin color

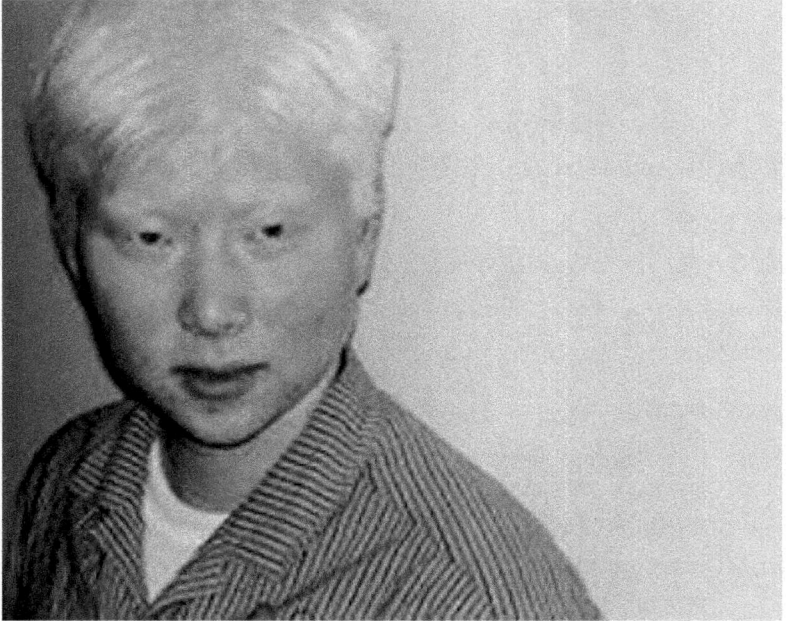

Figure 13-16 : An Asian man apparently effected by albinism of type 1 or 2

Figures 13-17 & 13-18 : The p gene of albinism can affect Caucasians of European descent as well

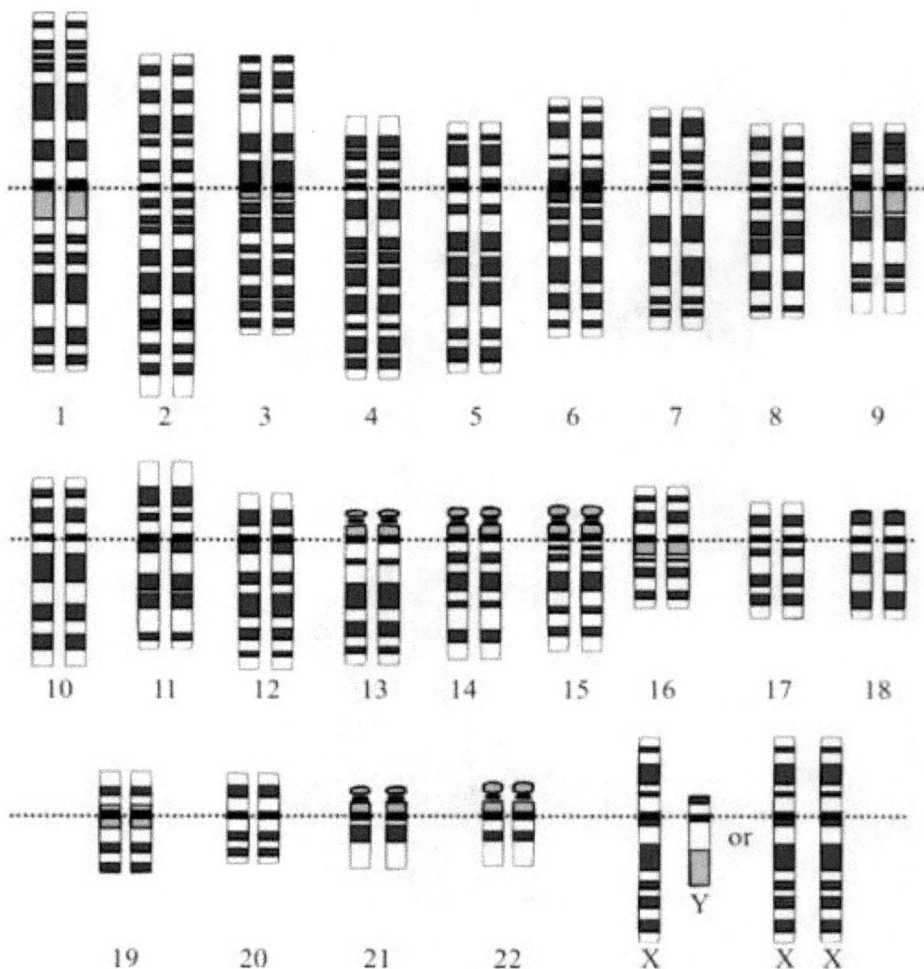

Figure 13-19 : The 23 Chromosomes that comprise the DNA
Double Helix Strand – Karyotype (Ideal Human Karotype)

13. 5 Genetic Drift

Current patterns of genetic variation reflect the history of migration and population growth in the human species. Some of the genetic variation among human groups reflects the history of the peoples of the Earth. The pattern of genetic variation in mitochondrial DNA indicates that the human species underwent a worldwide population expansion about 100,000 years ago. This was only one of several expansions of the world's population. The invention of agriculture led to expansions of farming peoples into Europe, eastern Asia, Oceania (the Pacific Islands groups of Polynesia, Melanesia, and Micronesia), and central Africa between 4000 and 1000 years ago; the domestication of the horse and associated military innovations led to several expansions of peoples living in the steppes of central Asia between 3000 and 500 years ago; and improvements in ships, navigation, and military organization led to the expansion of European populations during the last 500 years.(Boyd, Silk)

13.6 Phylogenetics – Reconstructing The Tree Of Life

In biology, phylogenetics is the study of evolutionary relationships among groups of organisms (e.g., species, populations), which are discovered through molecular sequencing data and morphological data matrices. The term phylogenetics derives from the Greek terms phylé (υυλή) and phylon (υῦλον), denoting "tribe", "clan", "race" and the adjectival form, genetikós (γενετικός), of the word genesis (γένεσις) "origin", "source", "birth". The result of phylogenetic studies is a hypothesis about the evolutionary history of taxonomic groups: their phylogeny. Evolution is regarded as a branching process, whereby populations are altered over time and may split into separate branches, hybridize together, or terminate by extinction. This may be visualized in a phylogenetic tree, a hypothesis of the order in which evolutionary events are assumed to have occurred. Phylogenetic analyses have become essential in researching the evolutionary tree of life.

The overall goal of National Science Foundation's Assembling the Tree of Life activity (AToL) is to resolve evolutionary relationships for large groups of organisms throughout the history of life, with the research often involving large teams working across institutions and disciplines.

Investigators are typically supported for projects in data acquisition, analysis, algorithm development and dissemination in computational phylogenetics and phyloinformatics. For example, RedToL aims at reconstructing the Red Algal Tree of Life.

Taxonomy, the classification, identification, and naming of organisms, is usually richly informed by phylogenetics, but remains methodologically and logically distinct. The degree to which taxonomy depends on phylogenies differs between schools of taxonomy: numerical taxonomy ignored phylogeny altogether, trying to represent the similarity between organisms instead; phylogenetic systematics tries to reproduce phylogeny in its classification without loss of information; evolutionary taxonomy tries to find a compromise between them in order to represent stages of evolution.

13.7 Why Reconstruct Phylogenies?

Phylogenetic reconstruction plays three important roles in the study of organic evolution. We have seen that descent with modification explains the hierarchical structure of the living world. Since new species always evolve from existing species, and species are reproductively isolated, all living organisms can be placed on a single phylogenetic tree, which we can then use to trace the ancestry of all living species.

In the remainder of this chapter, we will see how the pattern of similarities and differences observed in living things can be used to construct phylogenies and help to establish the evolutionary history of life.

There are several reasons why reconstructing phylogenies plays an important role in the study of evolution:

1. Phylogeny is the basis for the identification and classification of organisms. In the latter part of this chapter, we will see how scientists use phylogentic relationships to name organisms and arrange them into hierarchies. This endeavor is called taxonomy.

2. Knowing phylogentic relationships often helps to explain why species evolve certain adaptations and not others. Natural selection creates new species modifying existing body structures to perform new functions. To understand why a new organism evolved a particular trait, it helps to know what kind of organism it evolved from, and this is what phylogenetic trees tell us. The phylogenetic relationships among the apes provide a good example of this point. Most scientists used to believe that chimpanzees and gorillas shared a more recent common ancestor than either of them shared with humans. This view influenced their interpretation of the evolution of locomotion, or forms of movement, among the apes. All of the great apes are quadrupedal, which means that they walk on their hands and feet. However, gorillas and chimpanzees curl their fingers over their palms and bear weight on their knuckles, a form of locomotion called knuckle walking, while orangutans bear weight on their palms. Humans, of course, stand upright on two legs. Knuckle walking involves distinctive

3. modifications of the anatomy of the hand, and since human hands show none of these anatomical features, most scientists believed that humans dis not evolve from a knuckle-walking species. Because both chimpanzees and gorillas are knuckle walkers, it was generally assumed that this trait evolved in their common ancestor. However, recent measurements of genetic similarity now lead many scientists to believe that humans and chimpanzees are more closely related to one another than either species is to the gorilla. If this is correct, then the old account of the evolution of locomotion in apes must be wrong. Two accounts are consistent with the new phylogeny. It is possible that the common ancestor of

humans, chimpanzees, and gorillas was a knuckle walker and that knuckle walking was retained in the common ancestor of humans and chimpanzees. This would mean that knuckle walking evolved only once and that humans are descended from a knuckle-walking species.

Alternatively, the common ancestor of humans, chimpanzees, and other apes may not have been a knuckle walker. If this was the case, then knuckle walking evolved independently in chimpanzees and gorillas. Each of these scenarios raises interesting questions about the evolution of locomotion in humans and apes. If humans evolved from a knuckle walker, then perhaps more careful study will reveal the traces of our former mode of adaptation. If chimpanzees and gorillas evolved knuckle walking independently, then a close examination should reveal subtle differences in their morphology.

We can deduce the function of morphological features or behaviors by comparing the traits of different species. This technique is called the comparative method. Some scientists have argued that terrestrial (ground-dwelling) primates live in larger groups than arboreal (tree-dwelling) primates because terrestrial species are more vulnerable to predators and animals are safer in larger groups. To test the relationship between group size and terrestiality using the comparative method, we would collect data on group size and lifestyle (arboreal/terrestrial) for man different primate species. However, more biologists believe that only independently evolved cases should be contend comparative analyses, so we must take the phylogenetic relationships among species into account.

For many years, scientists constructed phylogenies only for the purpose of classification. The terms taxonomy and systematics were used interchangeably to refer to the construction of phylogenies and to the use of such phylogenies for naming and classifying organisms. With the recent realization that phylogenies have other important uses like the ones just described, there is a need for terms that distinguish phylogenetic construction from classification.

Here we adopt the suggestion of University of Zurich anthropologist Robert Martin to employ the term systematics to refer to the construction of phylogenies, and the term taxonomy to mean the use of phylogenies in naming the classification. Although this distinction may not seem important now, it will become more relevant as we proceed.

13.8 Reconstructing Phylogenies using Genetic Distance Data

Molecular data can be used to construct phylogenies, as we did when referring to the structure of myoglobin to construct the phylogeny of humans, chickens, and duck-billed platypuses. Molecular data are useful because they provide a large number of easily comparable characters. However, molecular data are also used in a completely different way to reconstruct phylogenies. Instead of looking for shared derived characters, biologists compute a measure of overall genetic similarity, called the genetic distance, between each pair of species. Scientists then reconstruct the phylogenies based on the assumption that pairs of genetically similar organisms have a more recent common ancestor than pairs of species less similar genetically.

The molecular technique called DNA hybridization is an important example of how genetic distance measures can be used to reconstruct phylogenetic relationships. In this process, an investigator takes tissue from a pair of organisms, extracts the DNA, and cuts into small fragments to bond to each other. When the DNA sequences from the two species are similar, any bonds created will be strong; when the DNA sequences are very different, the bonds will be relatively weak. Finally, the researcher gradually raises the temperature until the bonded pairs of DNA dissociate. The more strongly bonded (and hence similar) the molecules are, the higher the temperature needed to separate them.

Specifically, the temperature at which half of the strands from the two species and provides a measure of the genetic distance between the organisms. The investigator repeats this process for all of the pairs of species in the taxa she wants to classify. The genetic distance data can be used to construct a phylogeny. Other genetic distance data can be use molecular methods to assess the genotypes of DNA sequences of different species, and then the genetic distance is computed using these data.

* Phylogenetic reconstruction can also be based on genetic distance.

* These data are consistent with the hypothesis that genetic distance, as measured by DNA hybridization, changes at approximately a constant rate.

Rearranging the rows and columns if genetic distance data so that the most closely related species are next to each other produces a matrix. Gorillas, humans, chimpanzees, and orangutans are essentially the same genetic distance from gibbons. This is evidence that genetic distance changes at an approximately constant rate. To see why, think of genetic distance accumulating along the path leading form the last common ancestor of all of the hominoids to each living species. The total distance between any pair of living species is the sum of their paths from the last common ancestor. For example, the genetic distance between humans and gibbons is the accumulated distance between the last common ancestor of all hominoids and humans is the same as the elapsed since the branching off from the last common ancestor is the same for all hominoids, it follows that the rate of change of genetic distance along both of these paths through the phylogeny must be approximately the same, and that the genetic distance between two species is a measure of the time elapsed since both had a common ancestor. Evolutionists refer to genetic distances with a constant rate of change as molecular clocks because genetic change acts like a clock that measures time since two species shared a common ancestor.

Data from other groups or organisms suggest that genetic distances based on DNA hybridization and other methods often have this clock-like property, but there are also important exceptions. [111]

* DNA hybridization - A very simplistic explanation of DNA hybridization is when a single-stranded DNA (ssDNA) molecule bonds with a complementary ssDNA molecule from another source forming a "hybrid".

13.9 DNA Hybridization

DNA hybridization generally refers to a molecular biology technique that measures the degree of genetic similarity between pools of DNA sequences. It is usually used to determine the genetic distance between two species. When several species are compared that way, the similarity values allow the species to be arranged in a phylogenetic tree; it is therefore one possible approach to carrying out molecular systematics.

Charles Sibley and Jon Ahlquist, pioneers of the technique, used DNA-DNA hybridization to examine the phylogenetic relationships of avians (the Sibley-Ahlquist taxonomy) and primates.[1][2] Critics argue that the technique is inaccurate for comparison of closely related species, as any attempt to measure differences between orthologous sequences between organisms is overwhelmed by the hybridization of paralogous sequences within an organism's genome.

DNA sequencing and computational comparisons of sequences is now generally the method for determining genetic distance, although the technique is still used in microbiology to help identify bacteria.

The DNA of one organism is labeled, then mixed with the unlabeled DNA to be compared against. The mixture is incubated to allow DNA strands to dissociate and reanneal, forming hybrid double-stranded DNA.

Hybridized sequences with a high degree of similarity will bind more firmly, and require more energy to separate them: i.e. they separate when heated at a higher temperature than dissimilar sequences, a process known as "DNA melting".

To assess the melting profile of the hybridized DNA, the double stranded DNA is bound to a column and the mixture is heated in small steps. At each step, the column is washed; sequences that melt become single-stranded and wash off the column. The temperatures at which labeled DNA comes off the column reflects the amount of similarity between sequences (and the self-hybridization sample serves as a control). These results are combined to determine the degree of genetic similarity between organisms.

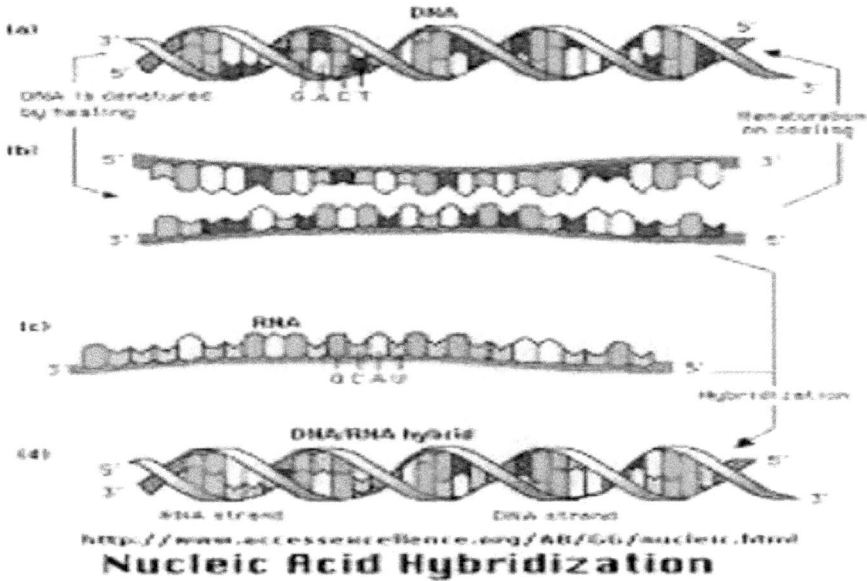

Figure 13-20 : Nucleic Acid Hybridization

13.10 Genetic Distance

Genetic distance refers to the genetic divergence between species or between populations within a species. It is measured by a variety of parameters. Smaller genetic distances indicate a close genetic relationship whereas large genetic distances indicate a more distant genetic relationship. The genetic distance can be used to compare the genetic similarity between different species, such as humans and chimpanzees. Within a spe,cies genetic distance can be used to measure the divergence between different sub-species. In its simplest form, the genetic distance between two populations is the difference in frequencies of a trait. For example, the frequency of Rh-negative individuals is 50.4% among Basques, 41.2% in France and 41.1% in England.

Thus, the genetic difference between the Basques and French is 9.2% and the genetic difference between the French and the English is 0.1% for the RH negative trait. The genetic distance of several individual traits can then be averaged to compute an overall genetic distance.

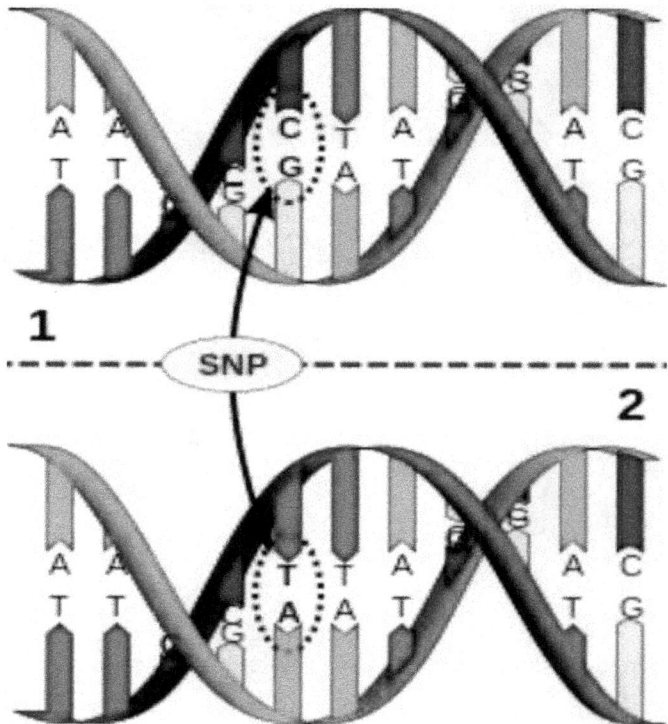

Figure 13-21 : DNA molecule 1 differs from DNA molecule 2 at a single base-pair location (a C/T polymorphism) giving rise to a genetic mutation or marker.

13.11 Human Genetic Variation

Human genetic variation is the genetic differences both within and among populations. There may be multiple variants of any given gene in the human population (genes), leading to polymorphism. Many genes are not polymorphic, meaning that only a single allele is present in the population: the gene is then said to be fixed.

No two humans are genetically identical. Even monozygotic twins, who develop from one zygote, have infrequent genetic differences due to mutations occurring during development and gene copy number variation. Differences between individuals, even closely related individuals, are the key to techniques such as genetic fingerprinting. Alleles occur at different frequencies in different human populations, with populations that are more geographically and ancestrally remote tending to differ more.

Causes of differences between individuals include the exchange of genes during meiosis and various mutational events. There are at least two reasons why genetic variation exists between populations. Natural selection may confer an adaptive advantage to individuals in a specific environment if an allele provides a competitive advantage. Alleles under selection are likely to occur only in those geographic regions where they confer an advantage. The second main cause of genetic variation is due to the high degree of neutrality of most mutations. Most mutations do not appear to have any selective effect one way or the other on the organism. The main cause is genetic drift, this is the effect of random changes in the gene pool. In humans, founder effect and past small population size (increasing the likelihood of genetic drift) may have had an important influence in neutral differences between populations. The theory that humans recently migrated out of Africa supports this.

The study of human genetic variation has both evolutionary significance and medical applications. It can help scientists understand ancient human population migrations as well as how different human groups are biologically related to one another. (Wiki)

For medicine, study of human genetic variation may be important because some disease-causing alleles occur more often in people from specific geographic regions. New findings show that each human has on average 60 new mutations compared to their parents. Apart from mutations, many genes that may have aided humans in ancient times plague humans today. For example, it is suspected that genes that allow humans to more efficiently process food are those that make people susceptible to obesity and diabetes today.

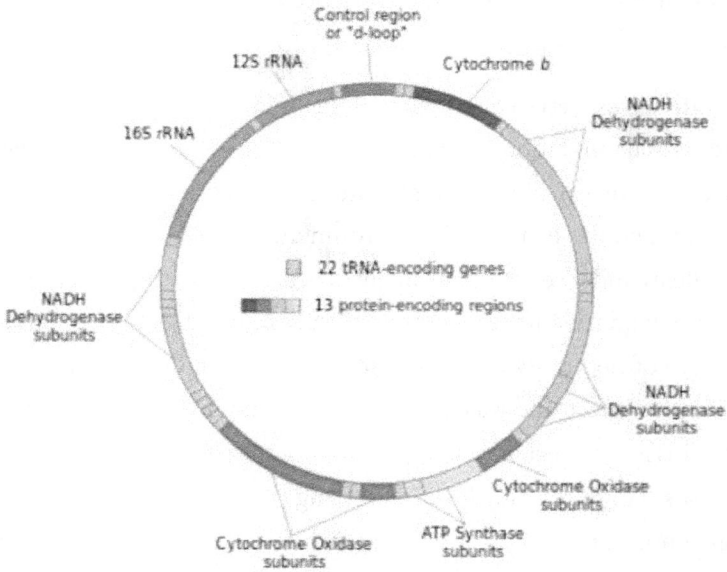

Figure 13-22 : A Mitochondrial DNA diagram

316

13.12 Genetic Variation – Measures of Variation

Genetic variation - variation in the alleles of genes - occurs both within and among populations. Since genetic variation provides the "raw material" for natural selection, it is important.

Genetic variation among humans occurs on many scales, from gross alterations in the human karyotype to single nucleotide changes.

Nucleotide diversity is the average proportion of nucleotides that differ between two individuals. The human nucleotide diversity is estimated to be 0.1% to 0.4% of base pairs. A difference of 1 in 1,000 amounts to approximately 3 million nucleotide differences, because the human genome has about 3 billion nucleotides.

13.13 Single Nucleotide Polymorphisms

A single nucleotide polymorphism (SNP) is difference in a single nucleotide between members of one species that occurs in at least 1% of the population. It is estimated that there are 10 to 30 million SNPs in humans.

SNPs are the most common type of sequence variation, estimated to comprise 90% of all sequence variations. Other sequence variations are single base exchanges, deletions, and insertions. SNPs occur on average about every 100 to 300 bases and so are the major source of heterogeneity.

A functional, or non-synonymous, SNP is one that affects some factor such as gene splicing or messenger RNA, and so causes a phenotypic difference between members of the species. About 3% to 5% of human SNPs are functional (see International HapMap Project). Neutral, or synonymous SNPs are still useful as genetic markers in genome-wide association studies, because of their sheer number and the stable inheritance over generations.

A coding SNP is one that occurs inside a gene. There are 105 Human Reference SNPs that result in premature stop codons in 103 genes. This corresponds to 0.5% of coding SNPs. They occur due to segmental duplication in the genome. These SNPs result in loss of protein, yet all these SNP alleles are common and are not purified in negative selection.

13.13 Single Nucleotide Polymorphisms

Structural Variation

Structural variation is the variation in the structure of an organism's chromosome. Structural variations, such as copy-number variation and deletions, inversions, insertions, and duplications, account for much more human genetic variation than single nucleotide diversity. This was concluded in 2007 from an analysis of the diploid full sequences of the genomes of two humans: Craig Venter and James D. Watson. This added to the two haploid sequences which were amalgamations of sequences from many individuals, published by the Human Genome Project and Celera Genomics respectively.

Copy Number Variation

A copy-number variation (CNV) is a difference in the genome due to deleting or duplicating large regions of DNA on some chromosome. It is estimated that 0.4% of the genomes of unrelated humans differ with respect to copy numbers. When copy-number variation is included, human-to-human genetic variation is estimated to be at least 0.5% (99.5% similarity). Copy number variations are inherited but can also arise during development.

Epigenetics

Epigenetic variation is variation in the chemical tags that attach to DNA and affect how genes get read. The tags, "called epigenetic markings, act as switches that control how genes can be read." At some alleles, the epigenetic state of the DNA, and associated phenotype, can be inherited across generations of individuals.

Genetic Variability

Genetic variability is a measure of the tendency of individual genotypes in a population to vary (become different) from one another. Variability is different from genetic diversity, which is the amount of variation seen in a particular population. The variability of a trait is how much that trait tends to vary in response to environmental and genetic influences.

13.13 Single Nucleotide Polymorphisms

Clines

In biology, a cline is a continuum of species, populations, races, varieties, or forms of organisms that exhibit gradual phenotypic and/or genetic differences over a geographical area, typically as a result of environmental heterogeneity. In the scientific study of human genetic variation, a gene cline can be rigorously defined and subjected to quantitative metrics.

Haplogroups

In the study of molecular evolution, a haplogroup is a group of similar haplotypes that share a common ancestor with a single nucleotide polymorphism (SNP) mutation. Haplogroups pertain to deep ancestral origins dating back thousands of years.

The most commonly studied human haplogroups are Ychromosome (Y-DNA) haplogroups and mitochondrial DNA (tuna) haplogroups, both of which can be used to define genetic populations. Y-DNA is passed solely along the patrilineal line, from father to son, while mtDNA is passed down the matrilineal line, from mother to both daughter and son. The Y-DNA and mtDNA may change by chance mutation at each generation.

A variable number tandem repeat (VNTR) is the variation of length of a tandem repeat. A tandem repeat is the adjacent repetition of a short nucleotide sequence. Tandem repeats exist on many chromosomes, and their length varies between individuals. Each variant acts as an inheritedallele, so they are used for personal or parental identification. Their analysis is useful in genetics and biology research, forensics, and DNA fingerprinting.

> *Short tandem repeats (about 5 base pairs) are called microsatellites, while longer ones are called minisatellites.*

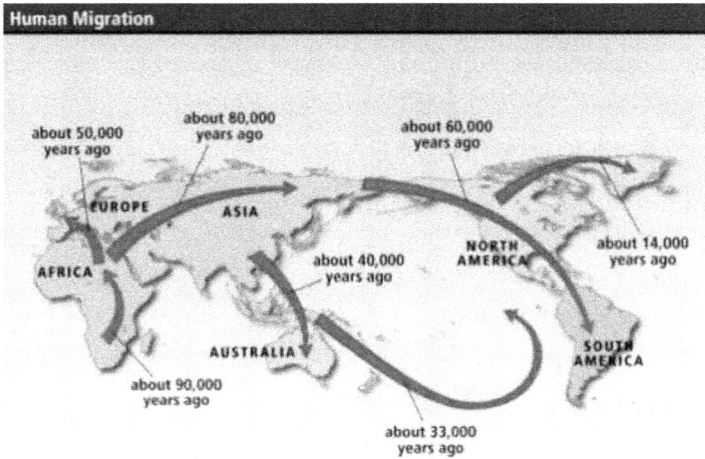

Figure 13-23 : Map of the migration of modern humans out of Africa, based on mitochondrial DNA.

13.14 History and Geographic Distribution

The Out of Africa theory (more precisely called "recent African origin of modern humans") is the most widely accepted explanation of the origin and early dispersal of anatomically modern humans, Homo sapiens sapiens. The theory states that archaic Homo sapiens evolved into modern humans solely in Africa, 200,000 to 100,000 years ago; around that time, one African subpopulation speciated when gene flow was restricted between African and Eurasian human populations; members of that subpopulation left Africa by 60,000 years ago and over time replaced earlier human populations such as Neanderthals and Homo erectus on Earth. Alternative theories include the multiregional origin of modern humans' hypothesis.

13.14 History and Geographic Distribution

The theory is supported by both genetic and fossil evidence. The hypothesis originated in the 19th century, with Darwin's Descent of Man, but remained speculative until the 1980s when it was supported by study of present-day mitochondrial DNA, combined with evidence from physical anthropology of archaic specimens. A large study published in 2009 found that modern humans probably originated near the border of Namibia and South Africa (reported as Namibia and Angola by BBC), and left Africa through East Africa; Africa contains the most human genetic diversity anywhere on Earth, and the genetic structure of Africans traces to 14 ancestral population clusters that correlate with ethnicity and culture or language. The study lasted 10-years and analyzed variations at 1,327 DNA markers of 121 African populations, 4 African American populations, and 60 non-African populations.

According to a 2000 study of Y-chromosome sequence variation, human Y-chromosomes trace ancestry to Africa, and the descendants of the derived lineage left Africa and eventually were replaced by archaic human Y-chromosomes in Eurasia. The study also shows that a minority of contemporary East Africans and Khoisan are the descendants of the most ancestral patrilineages of anatomically modern humans that left Africa 35,000 to 89,000 years ago. Other evidence supporting the theory is that variations in skull measurements decrease with distance from Africa at the same rate as the decrease in genetic diversity. Human genetic diversity decreases in native populations with migratory distance from Africa, and this is thought to be due to bottlenecks during human migration, which are events that temporarily reduce population size.

Population Genetics

In the field of population genetics, it is believed that the distribution of neutral polymorphisms among contemporary humans reflects human demographic history. It has been theorized that humans passed through a population bottleneck before a rapid expansion coinciding with migrations out of Africa leading to an African-Eurasian divergence around 100,000 years ago (ca. 5,000 generations), followed by a European-Asian divergence about 40,000 years ago (ca. 2,000 generations).Richard G. Klein, Nicholas Wade and Spencer Wells, among others, have postulated that modern humans did not leave Africa and successfully colonize the rest of the world until as recently as 60,000 - 50,000 years B.P., pushing back the dates for subsequent population splits as well.

The rapid expansion of a previously small population has two important effects on the distribution of genetic variation. First, the so-called founder effect occurs when founder populations bring only a subset of the genetic variation from their ancestral population. Second, as founders become more geographically separated, the probability that two individuals from different founder populations will mate becomes smaller. The effect of this assortative mating is to reduce gene flow between geographical groups, and to increase the genetic distance between groups. The expansion of humans from Africa affected the distribution of genetic variation in two other ways. First, smaller (founder) populations experience greater genetic drift because of increased fluctuations in neutral polymorphisms. Second, new polymorphisms that arose in one group were less likely to be transmitted to other groups as gene flow was restricted.

Our history as a species also has left genetic signals in regional populations. For example, in addition to having higher levels of genetic diversity, populations in Africa tend to have lower amounts of linkage disequilibrium than do populations outside Africa, partly because of the larger size of human populations in Africa over the course of human history and partly because the number of modern humans who left Africa to colonize the rest of the world appears to have been relatively low (Gabriel et al. 2002). In contrast, populations that have undergone dramatic size reductions or rapid expansions in the past and populations formed by the mixture of previously separate ancestral groups can have unusually high levels of linkage disequilibrium (Nordborg and Tavare 2002).

Many other geographic, climatic, and historical factors have contributed to the patterns of human genetic variation seen in the world today. For example, population processes associated with colonization, periods of geographic isolation, socially reinforced endogamy, and natural selection all have affected allele frequencies in certain populations (Jorde et al. 2000b; Bamshad and Wooding 2003). In general, however, the recency of our common ancestry and continual gene flow among human groups have limited genetic differentiation in our species.

13.15 Distribution of Variation

The distribution of genetic variants within and among human populations are impossible to describe succinctly because of the difficulty of defining a "population," the clinal nature of variation, and heterogeneity across the genome (Long and Kittles 2003). In general, however, an average of 85% of genetic variation exists within local populations, ~7% is between local populations within the same continent, and ~8% of variation occurs between large groups living on different continents. (Lewontin 1972; Jorde et al. 2000a; Hinds et al. 2005). The recent African origin theory for humans would predict that in Africa there exists a great deal more diversity than elsewhere, and that diversity should decrease the further from Africa a population is sampled. Long and Kittles show that indeed, African populations contain about 100% of human genetic diversity, whereas in populations outside of Africa diversity is much reduced, for example in their population from New Guinea only about 70% of human variation is captured.

13.16 Phenotypic Variation

Sub-Saharan Africa has the most human genetic diversity and the same has been shown to hold true for phenotypic diversity. Phenotype is connected to genotype through gene expression. Genetic diversity decreases smoothly with migratory distance from that region, which many scientists believe to be the origin of modern humans, and that decrease is mirrored by a decrease in phenotypic variation. Skull measurements are an example of a physical attribute whose within-population variation decreases with distance from Africa.

The distribution of many physical traits resembles the distribution of genetic variation within and between human populations (American Association of Physical Anthropologists 1996; Keita and Kittles 1997). For example, ~90% of the variation in human head shapes occurs within continental groups, and ~10% separates groups, with a greater variability of head shape among individuals with recent African ancestors (Relethford 2002).

13.16 Phenotypic Variation

A prominent exception to the common distribution of physical characteristics within and among groups is skin color. Approximately 10% of the variance in skin color occurs within groups, and ~90% occurs between groups (Relethford 2002). This distribution of skin color and its geographic patterning — with people whose ancestors lived predominantly near the equator having darker skin than those with ancestors who lived predominantly in higher latitudes — indicate that this attribute has been under strong selective pressure. Darker skin appears to be strongly selected for in equatorial regions to prevent sunburn, skin cancer, the photolysis of folate, and damage to sweat glands (Sturm et al. 2001; Rees 2003).

A study published in 2007 found that 25% of genes showed different levels of gene expression between populations of European and Asian descent. The primary cause of this difference in gene expression was thought to be SNPs in gene regulatory regions of DNA. Another study published in 2007 found that approximately 83% of genes were expressed at different levels among individuals and about 17% between populations of European and African descent.

Phylogenetics - a discipline of evolutionary biology which seeks to accurately depict the evolutionary relationships among living and non-living taxa. Homology/homologs - when traits are similar due to shared ancestry (the trait was inherited from a shared ancestor).

Synapomorphy – a derived trait shared by taxa due to common ancestry. A derived homology. Analogy/analogs – when traits are similar due to convergent evolution; the traits were not inherited from a common ancestor.

Taxa- general term for a taxonomic group (e.g., species, genera, families). Sister taxa are those presented as most-closely related in a phylogeny.

Clade- a group in a phylogenetic tree which begins with a node (ancestor) and includes everything more distal to the node (all descendents of the ancestor).

Monophyletic group- a proper clade; that is, a group which contains a common ancestor and all descendents of that ancestor (and no non-descendents).

Paraphyletic group- a group which contains a common ancestor and some, but not all, descendents of that ancestor.

13.18 Phylogenetics and Phylogenies

Phylogenetics is a discipline that aims to determine the true evolutionary relationships among organisms (The Tree of Life). Evolutionary biologists use a type of diagram, called a cladogram, to represent these evolutionary relationships or phylogenies. Cladograms (or phylogenetic trees) are sequentially branching trees. In these diagrams, you can gain information of the temporal pattern of diversification and shared ancestry.

Nodes on the trees (where branches meet) represent the common (shared) ancestor of all taxa beyond the node. If two taxa share a closer node than either share with a third taxon, then they share a more recent ancestor (they are more-closely related). Cladograms usually do not contain information about absolute time (e.g., in millions of years), but phylogenetic trees can be drawn which do depict the timing of events.

Reference the cladograms below, and make sure you understand how the following terms relate to the cladograms: sister taxa, node, branch, ancestor, descendent, most recent common ancestor.

Classification of organisms into the hierarchical system you are familiar with (e.g., Kingdom, Phylum, Class, etc.) is what the field of Taxonomy focuses on. Evolutionary biologists today believe that classification should represent true evolutionary relationships among organisms. Therefore, phylogenies are widely used for classification, and understanding the accurate phylogenetic relationships of organisms is important.

Taxa are best classified according to monophyletic groups, or clades. In these monophyletic groups, an ancestor (node) and all descendents are included. For example, —mammals‖ are a monophyletic group when we include the most recent common ancestor (MRCA) of all known mammals. —Reptiles‖ are not a monophyletic group if we exclude birds, since the MRCA of all reptiles is also a common ancestor of birds. Without birds included, —reptiles‖ is a paraphyletic group. Groups are polyphyletic when they include multiple taxa, but not the common ancestors.

13.19 Determining Phylogenetic Relationships

Phylogenetic relationships are established by analyzing homologous traits. Homologous traits are traits which are similar in two taxa because of shared ancestry. In contrast, analogous traits are similar because of convergent evolution of the two traits rather than inheritance from a shared ancestor. If we want to determine the accurate phylogeny of taxa, we need to concentrate on traits which are similar due to shared ancestry while ignoring analogies. Because traits are constantly modified throughout evolutionary history, some components of a trait may be homologous, while others are analogous. For example, wings in birds and bats are analogous and due to convergent evolution. However, certain components of wings are homologous – e.g., finger and humerus bones. Homologies are considered derived or ancestral depending on what clade you are looking at. Derived homologies (synapomorphies=shared, derived characters) are new to a clade of interest (first seen in ancestor of clade). Ancestral homologies (symplesiomorphies=shared, ancestral characters) arose before the common ancestor of the clade.

When determining phylogenetic relationships, we look only at the derived homologies (synapomorphies).Synapomorphies which define clades are often included in cladograms, as seen in the examples below.

Deuterostomy is a synapomorphy (derived homology) of the clade containing chordates and echinoderms.

Multicellularity is a symplesiomorphy (ancestral homology) of the same clade.

- Hair is a derived homology of all mammals.
- Having a backbone is an ancestral homology of all mammals.
- A backbone is a derived homology for the vertebrate clade.

The Principle of Parsimony is employed when using homologies to make a phylogeny. This principle favors the hypothesis that requires the fewest or simplest assumptions to explain an observation. In phylogenetics, the principle of parsimony invokes the minimal number of evolutionary changes to infer phylogenetic relationships. For example, it is more parsimonious to infer that a vertebral column evolved only once in a common ancestor of all living vertebrates than to infer that it evolved multiple times, once for fish, once for amphibians, etc. The first option requires fewer evolutionary changes.

The Principle of Parsimony states that trait origination is much less likely than trait inheritance in an ancestor-descendent relationship. The single origination and subsequent evolutionary inheritance of a trait is more likely than multiple originations of the same trait. When determining a phylogeny, we look at a number of traits in our taxa which are 1) not similar, 2) similar due to homology, or 3) similar due to convergence (analogy). Parsimony is invoked to construct a phylogeny that minimizes the number of changes in these traits (maximizes homology and minimizes analogy). In essence, we minimize the number of —tick marks‖ we have to make on the cladogram (see —tick marks‖ in the synapomorphy examples above.

13.20 There Is A Single Phylogeny For All Of Life

Based on homologies seen in all life forms (e.g., nucleic acids, metabolic pathways), it is most parsimonious to conclude that all of life as we know it (currently and historically) shares a common ancestor.

Humans are a third chimp : DNA homology suggests that humans are more closely related to living chimps than either are to other living primates.

Birds are dinosaurs :Based on traits common with mammals (e.g., endothermy), it was long thought that birds were more closely related to mammals than to other vertebrates. When looking at various skull and hip characteristics, however, birds share more traits with certain lineages of dinosaurs. When evidence arose that some dinosaurs had beak-like traits and feathers, it became clear that these bird traits were all homologous with dinosaurs and that the clade of living birds was, in fact, within the clade of dinosaurs.

A cladogram of the phylogenetic relationships of dinosaurs and birds. A cladogram of all reptiles, including birds.

13.21 Origin of Traits

Phylogenetics is used to identify where in a phylogeny trait first evolved. Once we have established an accurate phylogeny, we can make —tick marks‖ on a cladogram to indicate where a trait first appeared. These marks are usually placed on the branch before the MRCA of all taxa which share the trait. For example, we know that whales and their relatives (cetaceans) all have long, torpedo-shaped bodies. The closest relatives of cetaceans, the artiodactyls (e.g., cows, deer) do not have this body shape. Therefore, we know that this body shape must have arisen on the branch between the common ancestor of cetaceans and artiodactyls and the MRCA of all known cetaceans.

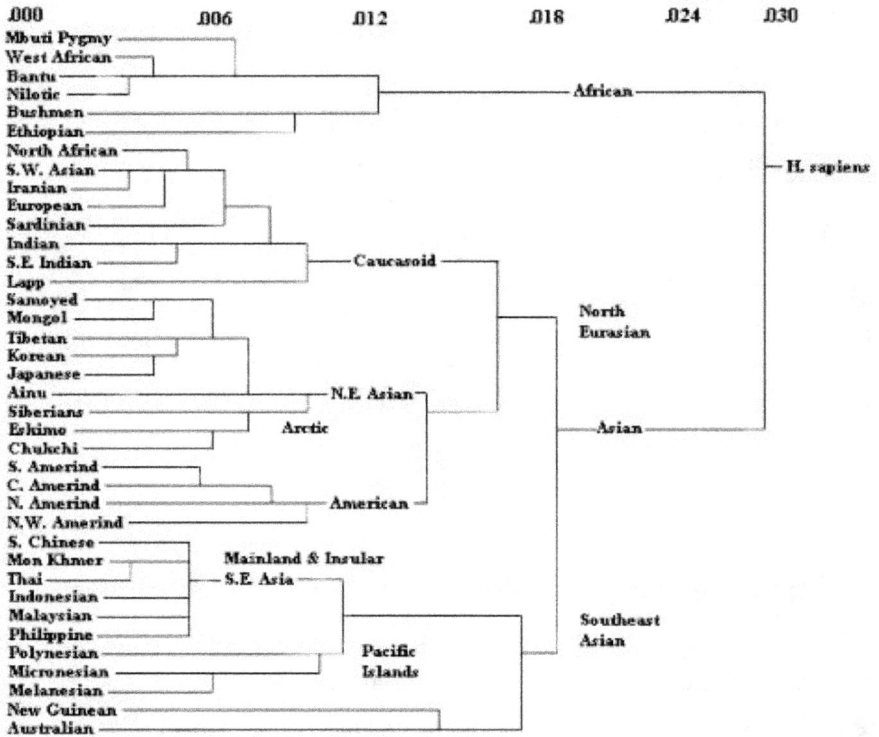

Figure 13-23 : Genetic Distances between Humans (genotype)

13.22 *Categorization of the World Population*

Race (classification of human beings) and Race and Genetics

New data on human genetic variation has reignited the debate about a possible biological basis for the categorization of humans into races. Most of the controversy surrounds the question of how to interpret the genetic data and whether conclusions based on it are sound. Some researchers argue that self-identified race can be used as an indicator of geographic ancestry for certain health risks and medications.

Although the genetic differences among human groups are relatively small, these differences in certain genes such as duffy, ABCC11, SLC24A5, called ancestry-informative markers (AIMs) nevertheless can be used to reliably situate many individuals within broad, geographically based groupings. For example, computer analyses of hundreds of polymorphic loci sampled in globally distributed populations have revealed the existence of genetic clustering that roughly is associated with groups that historically have occupied large continental and sub-continental regions (Rosenberg et al. 2002; Bamshad et al. 2003).

Some commentators have argued that these patterns of variation provide a biological justification for the use of traditional racial categories. They argue that the continental clusterings correspond roughly with the division of human beings into subSaharan Africans; Europeans, Western Asians, Central Asians, Southern Asians and Northern Africans; Eastern Asians, Southeast Asians, Polynesians and Native Americans; and other inhabitants of Oceania (Melanesians, Micronesians &Australian Aborigines) (Risch et al. 2002). Other observers disagree, saying that the same data undercut traditional notions of racial groups (King and Motulsky 2002; Calafell 2003; Tishkoff and Kidd 2004). They point out, for example, that major populations considered races or subgroups within races do not necessarily form their own clusters.

Furthermore, because human genetic variation is clinal, many individuals affiliate with two or more continental groups. Thus, the genetically based "bio-geographical ancestry" assigned to any given person generally will be broadly distributed and will be accompanied by sizable uncertainties (Pfaff et al. 2004).

In many parts of the world, groups have mixed in such a way that many individuals have relatively recent ancestors from widely separated regions. Although genetic analyses of large numbers of loci can produce estimates of the percentage of a person's ancestors coming from various continental populations (Shriver et al. 2003; Bamshad et al. 2004), these estimates may assume a false distinctiveness of the parental populations, since human groups have exchanged mates from local to continental scales throughout history (Cavalli-Sforza et al. 1994; Hoerder 2002). Even with large numbers of markers, information for estimating admixture proportions of individuals or groups is limited and estimates typically will have wide confidence intervals (Pfaff et al. 2004).

X	Africans	Europeans	Asians
Europeans	**36.5**		
Asians	**35.5**	**38.3**	
Indigenous Americans	**26.1**	**33.4**	**35**

Figure 13-24 : Human Genetic Clustering cell chart – Percentage similarity between two individuals from different clusters when 377 microsatellite markers are considered. (Boyd, Silk)

Genetic Clustering

Genetic data can be used to infer population structure and assign individuals to groups that often correspond with their self-identified geographical ancestry. Recently, Lynn Jorde and Steven Wooding argued that "Analysis of many loci now yields reasonably accurate estimates of genetic similarity among individuals, rather than populations. Clustering of individuals is correlated with geographic origin or ancestry."

Forensic Anthropology

Forensic anthropologists can determine geographic ancestry (i.e., Asian, African, or European) from skeletal remains with a high degree of accuracy by conducting bone analysis. Studies have shown that individual test methods such as mid-facial measurements and femur traits can be over 80% accurate, and in combination can be very accurate. However, the skeletons of mixed-ancestry individuals can exhibit characteristics of more than one ancestral group.

13.23 Admixture - Gene Flow

Gene flow between two populations reduces the average genetic distance between the populations. Genetic admixture is the result of interaction between well-differentiated populations. Admixture mapping is a technique used to study how genetic variants cause differences in disease rates between population. Recent admixture populations that trace their ancestry to multiple continents are well suited for identifying genes for traits and diseases that differ in prevalence between parental populations. African American populations have been the focus of numerous population genetic and admixture mapping studies, including studies of complex genetic traits such as white cell count, body-mass index, prostate cancer, and renal disease.

An analysis of phenotypic and genetic variation including skin color and socio-economic status was carried out in the population of Cape Verde which has a well-documented history of contact between Europeans and Africans. The studies showed that pattern of admixture in this population has been sex-biased and there is a significant interaction between socio economic status and skin color independent of the skin color and ancestry. Another study shows an increased risk of graft-versus-host disease complications after transplantation due to genetic variants in human leukocyte antigen (HLA) and non-HLA proteins.

Differences in allele frequencies contribute to group differences in the incidence of some monogenic diseases, and they may contribute to differences in the incidence of some common diseases (Risch et al. 2002; Burchard et al. 2003; Tate and Goldstein 2004). For the monogenic diseases, the frequency of causative alleles usually correlates best with ancestry, whether familial (for example, Ellisvan Creveld syndrome among the Pennsylvania Amish), ethnic (Tay-Sachs disease among Ashkenazi Jewish populations), or geographical (hemoglobinopathies among people with ancestors who lived in malarial regions).

To the extent that ancestry corresponds with racial or ethnic groups or subgroups, the incidence of monogenic diseases can differ between groups categorized by race or ethnicity, and health-care professionals typically take these patterns into account in making diagnoses.

Even with common diseases involving numerous genetic variants and environmental factors, investigators point to evidence suggesting the involvement of differentially distributed alleles with small to moderate effects. Frequently cited examples include hypertension (Douglas et al. 1996), diabetes (Gower et al. 2003), obesity (Fernandez et al. 2003), and prostate cancer (Platz et al. 2000). However, in none of these cases has allelic variation in a susceptibility gene been shown to account for a significant fraction of the difference in disease prevalence among groups, and the role of genetic factors in generating these differences remains uncertain (Mountain and Risch 2004).

Neil Risch of Stanford University has proposed that self-identified race/ethnic group could be a valid means of categorization in the USA for public health and policy considerations. While a 2002 paper by Noah Rosenberg's group makes a similar claim "The structure of human populations is relevant in various epidemiological contexts. As a result of variation in frequencies of both genetic and nongenetic risk factors, rates of disease and of such phenotypes as adverse drug response vary across populations. Further, information about a patient's population of origin might provide health care practitioners with information about risk when direct causes of disease are unknown."

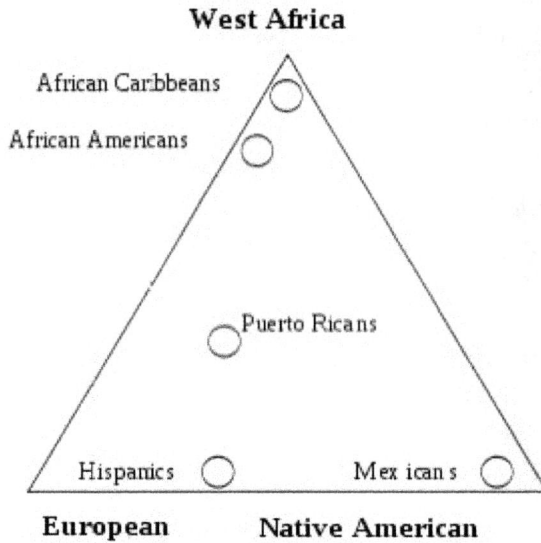

Figure 13-25 : The above triangle plot shows average admixture of five North American ethnic groups. Individuals that self-identify with each group can be found at many locations on the map, but on average groups tend to cluster differently.

13.24 *Genetic Variation Between Humans*

Genetic diversity often does not manifest itself as obvious differences in phenotype. Among humans, for example, there is more genetic diversity within Kalihari Bushmen than across all non-African populations combined. That's a result of the ancestors of the Bushmen having been in that area for hundreds of thousands of years, whereas all non-African populations passed through a common bottleneck about 50,000 years ago. (wiki)

Human Evolution Timeline:

Pre-Australopithecine

Pierolapithecus catalaunicus age: 11.9 myafossils found: nearly complete cranium, much of torso, arms and legs, parts of hands adaptations: small monkey-like hands, wider pelvis than predecessors, expanded rib cage for climbing location found: Spainnotes: may be common ancestor of chimpanzees and humans in the human evolution timeline; probably walked semi-upright using forelimbs.

Sahelanthropus Tchadensis

age: 6 – 7 myafossils found: nearly complete cranium, some fragments lower jaw & teeth brain size: 350 cc adaptations: ape-like, including small brain size, hominid features: brow ridges, small canine teeth location found: Chad, Central Africa notes: close to common ancestor of chimpanzees and humans in the human evolution timeline; unlikely to be bipedal.

Orrorin Tugenensis

age: c. 6 mya brain size: no cranium found fossils found: 13 fossils; partial femur, bits of jaw and teeth adaptations: unknown location found: Kenya, Africa notes: some indications of bipedality, but evidence is scant.

Ardipithecus Ramidus

age: 5.8 – 4.4 mya brain size: approx. 400 cc (chimp-size) fossils found: 110 specimens, almost complete skeleton adaptations: canine teeth intermediate between earlier apes & A. afarensis location found: Ethiopia notes: bipedal forest dweller(!), Canines smaller, not pointed.

Ardipithecus Kadabba

age: 5.8 – 5.2 mya brain size: no evidence fossils found: fragments, mostly teeth adaptations: smaller canines than apes (a feature of hominids) location found: Ethiopia notes:

Appendix I

Human Evolution Timeline: Genus Australopithecus

Australopithecus Anamensis

age: 4.2 – 3.9 mya brain size: unknown fossils found: tibia, lower jawbone adaptations: very likely bipedal, thick enamel on teeth (hominid characteristic) location found: Kenya, Africa notes: tibia shape indicates bipedalism

Australopithecus Afarensis

age: 3.9 – 2.9 mya brain size: 375 – 500 cc (large male–female size difference) fossils found: hundreds of fossils, fourteen partial skeletons, and footprints(!) adaptations: certainly bipedal, chimp-like skull, flat nose, no chin w/ human-like teeth, jaw between ape & human location found: Ethiopia, Tanzania, Cameroon notes: This is Lucy's species; she's an original, important "missing link."

Kenyanthropus Platyops

age: 3.5 – 3.3 mya brain size: unknown fossils found: two partial skulls adaptations: small ear canals, small teeth like Homo rudolfensis (branch of Homo habilis) location found: Lake Turkana, Kenya, Africa notes: scientists disagree whether these skulls deserve their own species.

Australopithecus Africanus

age: 3 – 2 mya brain size: 420 – 500 cc fossils found: partial skull, cranium, body & pelvis adaptations: fully human-shaped jaw, canine teeth reduced from afarensis, human-like pelvis location found: South Africa notes: everything seems clearly adapted from afarensis.

Australopithecus Garhi

age: 2.5 mya brain size: ? fossils found: partial skull adaptations: ? location found: Ethiopia notes: some evidence for tool use, otherwise not much known

Australopithecus Aethiopicus

age: 2.6 – 2.3 mya brain size: 410 cc fossils found: "The Black Skull" and some minor fossils adaptations: baffling; small brain, massive face & very strong jaw location found:

Human Evolution Timeline: Genus Australopithecus

Ethiopia, Kenya notes: some say this is perfect intermediate between afarensis and boisei, limited information

Australopithecus Robustus

age: 2 – 1.5 mya brain size: 530 cc fossils found: many, mostly cranial and dental adaptations: body like africanus, but larger skull, molars, and massive face & brow ridge, small canines location found: South Africa notes: bones found w/ robustus fossils may have been tools, this line of hominids had strong jaws and huge molars for crunching rough food, then probably went extinct (no descendants in southern Africa).

Australopithecus Sediba

age: 1.95 – 1.8 mya brain size: 420 cc fossils found: two partial skeletons, most of cranium in juvenile adaptations: more Homo adaptations than any other Australopithecine, pelvis is particularly advanced for bipedality so that it may have been capable of running, and brain case is shaped more towards human location found: South Africa notes: New discovery, described in April, 2010 (!); pelvis and leg indicate it's bipedal; more recent than many Homo fossils, thus unlikely to be human ancestor; probable adult height 4'6" for male, 4'2" for female.

Australopithecus Boisei

age: 2.1 – 1.1 mya brain size: 530 cc fossils found: many, mostly cranial and dental adaptations: may just be robustus location found: Tanzania, Kenya, & Ethiopia, Africa notes: used to be Zinjanthropus; an unlikely human ancestor, probably went extinct due to over-specialization with it's huge molars and small canines; some make them Paranthropus boisei.

Human Evolution Timeline: Genus Homo

Homo Gautengensis

age: 2 million to 600,000 years ago brain size: awaiting publication fossils found: partial skull, several jaws, teeth, and other bones adaptations: bipedalism, further information awaiting publication location found: Sterkfontein Caves, South Africa notes: Very stocky! 3 foot tall and 110 pounds; bipedal tree dweller.

Homo Habilis

age: 2.4 – 1.5 mya brain size: 500 – 800 cc fossils found: many adaptations: primitive face, smaller teeth than australopithecines, human-shaped brain, had "Broca's area," a section of brain we currently use for speech location found: Kenya & Tanzania, Africa notes: named habilis for tools found with fossils, may need to be more than one species, Homo rudolfensis is suggested in addition.

Homo Georgicus

age: 1.8 mya brain size: 600 – 730 cc fossils found: thirty or so partial skulls & a partial skeleton adaptations: intermediate between habilis and erectus location found: Dminisi, Georgia (eastern Europe) notes: First hominids out of Africa, a huge surprise because they didn't have the brains or tools of Homo erectus.

Homo Floresiensis

age: 1.1 mya to 17,000 years ago (possibly 2 mya!) brain size: 420 cc fossils found: several almost full skeletons adaptations: dwarf hominim found on island; large feet, ape-like hands, small brain, similar features to H. habilis location found: Flores, Indonesia notes: dwarves of species commonly develop on islands, these were 3 feet tall; hunted dwarf elephants and large rats; new research on relative brain size suggests they descended from H. georgicus or habilis rather than erectus.

Homo Erectus

age: 1.8 million – 300,000 years ago brain size: 750 – 1225 cc fossils found: lots and lots adaptations: jaw still protrudes, no chin, thick brow ridges, small forehead, but larger brains and excellent walkers location found: Africa, Europe, Asia notes: probably used fire, brains got larger over time.

Human Evolution Timeline: Genus Homo

Homo Ergaster

age: same as erectus brain size: slightly smaller than erectus fossils found these would be the African erectus fossils reclassified adaptations: taller & thinner & different-shaped brow ridges than European & Asian erectus fossils location found Africa notes: some scientists make the African erectus fossils to be Homo ergaster.

Homo Antecessor

age: 780,000 years ago, brain size: 1000 cc fossils found: six individuals' adaptations: mid-facial area modern, other parts primitive location found: Spain notes: oldest European hominids, may just be erectus or early heidelbergensis.

Homo sapiens (wise)or Homo Heidelbergensis

age: 500,000 years ago, brain size: 1200 cc fossils found: many skulls' adaptations: intermediate between erectus and human in thickness of bones & size of teeth, receding foreheads and chins location found: Europe notes: some scientists consider these the European version of Homo erectus; later ones are difficult to distinguish from early Homo sapiens; sometimes heidelbergensis is called Homo sapiens (archaic) in contrast to modern humans who are Homo sapiens sapiens.

Denisovans

(Sometimes referred to as Homo denisova, but this is not correct ... yet.) age: 500,000 – 30,000 years ago brain size: unknown fossils found: child's finger and adult molar (significant DNA sequenced) adaptations: unknown location found: Denisova cave, Siberia notes: Living humans, especially in the islands around New Guinea, inherited genes from this population. We know nothing about them other than they existed, and there is debate about whether they deserve their own species designation. The sequenced DNA has shed much light on the geographical spread of humans and human ancestors, but there is also much debate as the discovery is so new.

Human Evolution Timeline: Genus Homo

Homo Neanderthalensis or Homo sapiens Neanderthalis

(or just Neandertals) age: 400,000 – 30,000 years ago brain size: 1450 cc fossils found: lots and lots, entire genome sequenced adaptations: mid-facial area protrudes, long low braincase, thick & strong, near our height; many adaptations seem like adaptations for cold location found: Europe, Middle East, and Asia note: first hominids to bury dead, lots of tools & weapons lived brutal lives; the brain was larger than ours! Often spelled Neanderthal now. Humans carry some Neanderthal genes, indicating interbreeding. Thus, by some paleontologists' definition, they cannot be a separate species.

Cro-Magnon(Homo sapiens)

age: 50,000 – 10,000 years ago brain size: 1350 ccfossils found: many adaptations: These are now classified as anatomically-modern humans location found: Europenotes: Our ancestors in Europe were thicker in tooth and bone than we are.

Homo Sapien Sapien(Modern Humans)

age: 195,000 years ago – 2009 brain size: 1350 ccfossils found: living samples available in large quantities adaptations: forehead rises sharply, small, or non-existent brow ridges, prominent chin, thin skeleton location found: everywhere notes: tooth and face bone size is still decreasing over the last 20,000 years of the human evolution timeline!

Acheulean

a tool industry found at sites dated at 0.3 to 1.5 mya and associated with Homo ergaster and some Homo sapiens. Named after the French village of St. Acheul, where it was first discovered, the Acheulean industry is dominated by teardrop-shaped hand axes and blunt cleavers.

Achondroplasia

a genetic disease caused by a dominant gene that leads to the development of short stature and disproportionately short arms and legs.

Adaptation

a feature of an organism created by the process of natural selection.

Adaptive Grade

the basic way that an animal makes a living. Distantly related animals can belong to the same adaptive grade.

Adaptive Radiation

the process in which a single lineage diversifies into a number of species, each characterized by distinctive adaptations. The diversification of the mammals at the beginning of the Cenozoic era is an example of adaptive radiation.

Adenine

one of the four bases of the DNA molecule. The complementary base of adenine is thymine.

Affiliative

friendly.

Allele

one of two or more alternative forms of a gene. For example, the A and S alleles are two forms of the gene controlling the amino acid sequence of one of the subunits of hemoglobin.

Allopatric Speciation

speciation occurs when two or more populations of a single species are geographically isolated from each other and then diverge to form two or more new species.

Altruism

behavior that reduces the fitness if the individual performing the behavior (the actor) but increases the fitness of the individual affected by the behavior (the recipient)

Amino Acids

molecules that are linked in a chain to form proteins. There are 20 different amino acids that share the same molecular backbone but have a different side chain attached.

Analogy

(analogous, adj.) traits that are similar because if convergent evolution, not common descent. For example, the fact that humans and kangaroos are both bipedal is an analogy, not a homology.

Ancestral Trait

a trait that appears earlier in the evolution of a lineage or clade. Ancestral traits are contrasted with derived traits, which appear later in the evolution of a lineage or clade. For example, the presence of a tail is ancestral in the primate lineage, while the absence of a tail is derived. Systematists must avoid using ancestral similarities when constructing phylogenies.

Angiosperms

the flowering plants. The radiation of the angiosperms during the Cretaceous period may have played an important role in the evolution of the primates.

Antagonistic Pleiotropy Hypothesis

the basis of one explanation for the evolution of senescence. Antagonistic pleiotropy is the condition in which a gene has a positive effect on one component of fitness and a negative effect on some other component of fitness. For example, a gene might increase fecundity early in life but reduce the probability of survival later in life.

Anthropoid

a member of the primate suborder that includes the monkeys and apes. The only other primate suborder (the prosimians) includes lemurs, lorises, and tarsiers.

Anticodon

the sequence of bases on a transfer RNA molecule that binds complementarily to a particular codon. For example, the codon, ATC to the corresponding anticodon is TAG because A binds to T and G to C.

Archaeologist

a scientist who studies the material remains of past cultures and people.

Archaic Homo Sapiens

an older term for hominids with larger brains and more modern crania that appear in the fossil record about 500 kya in Africa and Europe, and somewhat later in eastern Asia.

Argon-argon

dating a sophisticated variant of the potassium-argon dating method that allows very small samples to be accurately dated.

Atlatl

a spear-thrower – a device that lengthens the arm and allows the spear to be thrown with much greater force.

Atresia

the decline in the number of viable germ cells that occurs as a woman gets older.

Attribution

the ability to form a mental model of the mind of another individual. For example, adult humans understand that other people may have different knowledge about the world than they do. Small children do not have this ability.

Basal Metabolic Rate

the rate of energy use required to maintain life when an animal is at rest.

Bases

one of four molecules – adenine, guanine, cytosine, and thymine – that are bound to the DNA backbone. Different sequences f bases encode the information necessary for protein synthesis.

Biface

a flat stone tool made by working both sides of a cre until there is an edge along the entire circumference.

Biochemical Pathways

the chains of chemical reactions by which organisms regulate their structure and chemistry.

Biological Species Concept

the concept that species are defined as a group of organisms that cannot interbreed in nature. Adherents of the biological species concept believe that the resulting lack of gene flow is necessary to maintain differences between closely related species.

Bipedal

walking upright on two (hind) legs.

Blades

stone tools made from flakes that are at least twice as long as they are wide. Blades dominate the tool traditions of the Uppr Paleolithic.

Blending Inheritance

a model of heritance, widely held during the 19th century, in which the hereditary material of the mother and father was thought to blend irreversibly together in the offspring.

Bottleneck

a server but temporary reduction in population size that reduces the amount of genetic variation present in the population.

Broca's area

a region of the human brain located in the left hemisphere. People who suffer damage to Broca's area often have difficulty with syntax.

Canalized

describes a trait that is very insensitive to environmental conditions during development, resulting in similar phenotypes in a wide range of environments.

Carbon-14 Dating

a dating method based on an unstable isotope of carbon with atomic weight of 14. Carbon 14 is produced in the atmosphere by cosmic radiation and is taken up by living organisms. After organisms die, the carbon-14 present in their bodies decays to stable isotope nitrogen-14 at a constant rate. By measuring the ratio of carbon-14 to the stable isotope of carbon (carbon-12) in organic remains, scientists can estimate the length of time that has passed since the organism died. Also called *radiocarbon dating*.

Carbohydrates

certain organic molecules with the formula $CnH2nn$ including common sugars and starches.

Catalysis

the process by which a catalyst increases the rate at which a chemical reaction occurs.

Character Displacement

the result of competition between two species that causes the members of different species to become morphologically or behaviorally more different from each other.

Chatelperronian

an Upper Paleolithic tool industry found in France and Spain that dates from 36 kya to 32 kya and is associated with Neanderthal fossil remains.

Chopper

a simple tool made by removing a few flakes from a stone to produce an edge. Choppers dominate the Oldowan tool industry.

Chromosomes

linear bodies in the cell nucleus that carry genes and that appear during cell division. Staining cells with dyes reveal those different chromosomes are marked by different banding patterns.

Cladistic Systematics

a system for classifying organisms in which patterns of decent are the only criteria used in classification.

Codon

a sequence of three DNA bases on a DNA molecule that comprises one —word‖ in the message used to create a specific protein. There are 64 different codons.

Coefficient of Relatedness

an index measuring the degree of genetic closeness between two individuals. The index ranges from 0 (for no relation) to 1 (which occurs only between and individual and itself, or between identical twins). For example, the coefficient of relatedness between and individual and its parents or its siblings is 0.5.

Cognitive Map

a mental representation of the location of objects in space and time that allows for efficient navigation.

Computer Tomography

an X-ray technique that generates three dimensional images.

Content-dependent Reasoning

modes of reasoning that depend on what is being reasoned about. The laws of formal logic are supposed to apply to any subject. However, humans seem to reason differently when given problems that are logically identical but deal with different subjects.

Continental Drift

the movement over the surface of the globe of immense plates of relatively light material that make up the continents.

Corpus Luteum

a mass of cells that forms in a woman's ovary after ovulation that in turn produces the hormone progesterone.

Correlated Response

an evolutionary change in one character caused by selection on a second, correlated character. For example, selection favoring only long legs will also increase are length if arm length and leg length are positively correlated.

Cranium

the braincase and bones of the face. Together the cranium and the lower jaw (mandible) make up the skull.

Creole

a grammatically completely new language that has arisen on plantations and in other situations in which people speaking many different languages live close together.

Cross

in genetics, a mating between chosen parents.

Crossing Over

the exchange of genetic material between homologous chromosomes during meiosis. Crossing over causes recombination of genes carried on the same chromosome.

Cytoplasm

the material inside the cell but outside of the nucleus.

Cytosine

one of the four bases of the DNA molecule. The complementary base of cytosine is guanine.

Deoxyribonucleic Acid

-DNA

Derived Trait

a trait that appears later in the evolution of a lineage or clade. Derived traits are contrasted with ancestral traits, which appear earlier in the evolution of a lineage or clade. For example, the absence of a tail is derived in the hominid lineage, while the presence of a tail is ancestral. Systematists seek to use derived similarities when constructing phylogenies.

Discontinuous Variation

phenotypic variation in which there are a discrete number of phenotypes with no intermediate types. Pea color in Mendel's experiments) the likelihood that the offspring will survive and reproduce.

Dizygotic Twins

that result from the fertilization of two separate ova by two separate sperm. Dizygotic twins are no more closely related than other full siblings. See also monozygotic twins.

DNA (deoxyribonucleic acid)

the molecule that carries hereditary information in almost all living organisms. It consists of two very long phosphate-sugar backbones to which the bases adenine, cytosine, guanine, and *thymine* are bound. Hydrogen bonds between the bases bind to two strands together.

DNA Hybridization

a laboratory technique in which the similarity in sequence between DNA taken from different sources is assessed by measuring the temperature at which the two types of DNA dissociate.

Dominance

the ability of one individual to intimidate or defeat another individual in a pairwise (dyadic) encounter. In some cases, dominance is assessed from the outcome of aggressive encounters, and in other cases dominance is assessed from the outcome of competitive encounters.

Dominance Hierarchy

a ranking of individuals in a group that reflects their relative dominance.

Dyadic

describes and interaction that involves two individuals. Also called pairwise.

Endangered Species

a species that is in danger of becoming extinct due to its small population size or destruction of its habitat.

Endocranial Volume

the volume inside the braincase.

Environmental Variation

phenotypic differences between individuals that exist because individuals developed in different environments.

Enzyme

a protein that serves as a catalyst. Enzymes can control the chemical composition of cells by causing some chemical reactions to occur much faster than others.

Equilibrium

a steady state in which either gen or genotypic frequencies do not change.

Eukaryotes

organisms whose cell have cellular organelles, cell nuclei, and chromosomes. All plants and animals are eukaryotes. See also prokaryotes.

Exons

segments of the DNA in eukaryotes that are translated int protein. Exons contrast with intervening introns, which are not translated into protein.

Family

a taxonomic level above genus but below order. Thus, a family may contain several genera, and an order may contain several families. Humans belong to the family Hominidae, and the other great apes belong to the family Pongidae.

Fitness

a measure of an individual's genetic contribution to subsequent generations.

Fixation

a state that occurs when all of the individuals in a population are homozygous for the same allele at a particular locus.

Foragers

people who subsist by gathering, fishing, and hunting. Until about 10 kya all humans were foragers. Foragers are also called hunter-gatherers.

Fossil

traces of life more than 10,000 years old preserved in rock. Fossils can be mineralized bones, plant parts, impressions of soft body parts, or tracks.

Gametes

in animals, eggs, and sperm.

Founder Effect

a form of genetic drift that occurs when a small population colonizes a new habitat and subsequently greatly increases in number. Random genetic changes due to the small size of the initial population are amplified by subsequent population growth.

Gene

a segment of the chromosome that produces a recognizable effect on phenotype and segregates as a unit during gamete formation.

Gene Flow

the movement of genes from one population to another or one part of a population to another, as the result of interbreeding.

Gene Frequency

that fraction of the genes at a genetic locus that are a particular allele (and thus the same as allele frequency). For example, in a population in which there are 250 AA individuals, 200 AS individuals and 50 SS individuals, there are 700 copies of the A allele and 300 copies of the S allele, and therefore the frequency of the S allele is 0.3.

Genetic Distance

a measure of the overall genetic similarity of individuals or species. The best estimates of genetic distance utilize large numbers of genes.

Genetic Drift

Random change in gene frequencies due to sampling variation that occur in any finite population. Genetic drift is more rapid in small populations than in large populations.

Genetic Markers

genes whose position on a chromosome is known and can be used to locate other nearby, linked genes.

Genetic Variation

phenotypic differences between individuals that result from the fact that they have inherited different genes from their parents.

Gene Tree

a phylogenetic tree tracing the pattern of descent for a particular gene.

Genotype

the combination of alleles that characterizes an individual at some set of genetic loci. For example, in populations with only the A and S alleles at the hemoglobin locus, there are three possible genotypes at that locus: AA, AS, and SS, (SA is the same as AS).

Genotypic Frequency

the fraction of individuals in a population who have a particular genotype.

Genome

all of the genetic information carried by the organism.

Genus

a taxonomic category below the family and above the species. There may be several species in a genus, and several genera (the plural of genus) in a family.

Germ Cells

cells in reproductive organs that give rise to gametes.

Gerontologist

a scientist who studies aging.

Gestational Hypertension

a form of hypertension (high blood pressure) that occurs during pregnancy. Gestational hypertension is associated with high birth weight and may be caused by manipulation of the mother's body by the fetus.

Glucose

a simple sugar that is used as a medium of energy transport in many animals.

Grooming

the process of picking through the hair to remove dirt, dead skin, ectoparasites, and other material, a common form of affiliative behavior among primates.

Guanine

one of the four bases of the DNA molecule. The complementary base of guanine is cytosine.

Grammar

all of the rules of meaning our brains use to interpret information provides by language.

Haft

to attach a spear point, ax head, or similar implement to a handle. Hafting greatly increases the force that can be applied to the tool.

Haploid

a cell with only one copy of each chromosome. Gametes are haploid, as are the cells of some asexual organisms.

Hardy-Weinbergequilibrium

the unchanging frequency of genotypes that results from sexual reproduction and in the absence of other evolutionary forces such as natural selection, mutation, or genetic drift.

HDC *(high digestive capacity)*

the ability to digest lactose in adulthood.

Hemoglobin

a protein in blood that carries oxygen, including two α (alpha) and to (beta) β subunits.

Heritability

the fraction of the phenotypic variation in the population that is the result of genetic variation.

Heterozygous

refers to a diploid organism whose cells carry two different alleles for a particular genetic locus. Organisms that are heterozygous are called —*"heterozygotes"*.

Homeotherms

organisms that maintain constant body temperature via internal physiological mechanisms.

Home Range

that area in which an individual or a group of animals travels, feeds, rests, and socializes. Territorial species actively defend the borders of their home ranges.

Hominids

any member of the family Hominoidea, which includes humans, all living apes, and numerous extinct ape and humanlike species from the Miocene, Pliocene, and Pleistocene epochs.

Homo Heidelbergensis

Middle Pleistocene hominids from Africa and western Eurasia. These hominids had large brains and very robust skulls and postcrania.

Homologous Chromosomes

a pair of chromosomes in a diploid cell in which one member of the pair is derived from the father and one is derived from the mother.

Homology *(homologous, adj.)*

traits that are similar because of common ancestry, not convergence. For example, the fact that gorillas and baboons are both quadrupedal is due to the fact that they ae both descended from a quadrupedal ancestor.

Homozygous

refers to a diploid organism whose chromosomes carry two copies of the same allele at a single genetic locus. Organisms that are homozygous are called — *"homozygotes"*.

Hormone

a chemical substance produced by the endocrine system that is transported through the bloodstream to another part of the body where it produces a particular physiological response.

Human Growth Hormone *(hGH)*

a hormone with a number of effects on growth and that blocks the effect of insulin.

Human Universals

features that characterize humans all over the world.

Humerus

the bone in the upper part of the forelimb (arm).

Hunter-Gatherers

people who subsist by gathering, fishing, and hunting. Until about 10 kya ago all humans were hunter-gatherers. Also called *foragers*.

Hybrid Zone

a geographical region where two or more populations of the same species or two different species overlap and interbreed. Hybrid zones usually occur at eh habitat margins of the respective populations.

Insulin

a substance that is created by the pancreas and is involved in the regulation of blood sugar.

Isotope

a chemical element with the same atomic number and properties as another element but having a different atomic weight. Unstable isotopes spontaneously change into more stable isotopes. Also see radioactive decay.

Knuckle Walking

a form of quadrupedal locomotion in which, in the forelimbs, weight is supported by the knuckles, rather than the palm or outstretched fingers. Chimpanzees and gorillas and knuckle walkers.

Linkage

the tendency for alleles at certain loci to be transmitted to the same gamete as a unit because they are located on the same chromosome. The closer together two loci are, the more likely they are to be linked.

Locus *(loci, pl)*

the position on a chromosome that is occupied by a particular gene.

Macroevolution

evolution of a new species, families, and higher taxa.

Mating Guarding

a form of mating in which the male defends his mate after copulating to prevent other males from mating with her.

Mating Systems

forms of courtship, mating, and parenting behavior that characterize particular species or populations. An example is polygyny.

Matrilineage

individuals related through the maternal line.

Maxilla

the upper jaw.

Meiosis

the process of cell division in which haploid gametes (eggs and sperm) are created.

Mendel's Principles

the principles derived from Gregor Mendel's plant-breeding experiments. The first principle states that the observed characteristics of organisms are jointly determine by two particles (genes), one inherited from the mother and one inherited from the father. The second principle states that each of these genes is equally likely to be transmitted when gametes are formed.

Menopause

the time in a woman's reproductive career when menstrual cycles stop, usually around the age of 50 in the United States.

Messenger RNA (mRNA)

a form of RNA that carries specifications for protein synthesis from DNA to the ribosomes.

Microevolution

evolution of populations within a species.

Middle Pleistocene

the period for 900 kya to 137 kya. The Pleistocene is divided into three parts: the Lower, Middle, and Upper Pleistocenes. The beginning of the Middle Pleistocene coincides with the initial growth of continental glaciers in Europe and ends with the termination of the next-to-last glacial period.

Middle Stone Age (MSA)

the stone tool industries of subSaharan Africa and southern and eastern Asia that existed 150 kya to 30 kya. The SA is the counterpart of the Middle Paleolithic (Mousterian) in Europe. The MSS industries varied, but flake tools were manufactured in all of them.

Minor Marriage

a form of marriage, formerly widespread in China, in which children were betrothed in infancy and then raised together in the household of the prospective groom.

Mitochondria (Mitochondrion, Sing)

cellular organells involved in basic energy processing. See also mitochondrial DNA.

Mitosis

the process of division of somatic (normal body) cells through which new diploid cells are created.

Modern Homo Sapiens

Human beings that first appear in the fossil record 100 kya.

Modern Synthesis

an explanation for the evolution of continuously varying traits that combines the theory and empirical evidence of both Mendelian genetics and Darwinism.

Molecular Clock

the hypothesis that genetic change occurs at a constant rate and thus can be used to measure the time elapsed since two species shared a common ancestor. It is based on observed regularities in the rate of genetic change along different phylogenetic lines.

Monozygotic Twins

twins that result from the fertilization of one ova by a single sperm. Early in development the fertilized egg splits to create two zygotes. See also dizygotic twins.

Morphology

1) the form and structure of an organism; also a field of study that focuses on the form and structure of organisms. 2) A part of grammar that governs the way that words are put together.

mRNA

See messenger RNA.

mtDNA

See mitochrondrial DNA.

Multiregional Hypothesis

a hypothesis of the evolution of anatomically modern humans that holds that populations of archaic Homo sapiens throughout the world were linked to gene flow.; This allowed humans to evolve from Homo erectus into modern Homo sapiens as a single species all over the Old World. However, there was considerable regional variation in morphology among populations.

Mutation

a spontaneous change in the chemical structure of DNA.

Mutation Accumulation

hypothesis a nonadaptive theory of aging that considers aging the result of the accumulation of mutations that affect only older individuals.

Mutualism(mutualistic, adj.)

behavior that increases the fitness of both actor and recipient.

Natal Group

the group into which an individual is born. In many primate species the females remain in their natal groups throughout their lives, while the males emigrate and join new groups.

Natural Selection

the process that produces adaptation. Natural selection is based on three postulates: 1) the availability of resources is limited; 2) organisms vary in the ability to survive and reproduce; and 3) traits that influence survival and reproduction are transmitted from parents to offspring. When these three postulates hold, natural selection produces adaption.

Neanderthals

a form of archaic Homo sapiens found in western Eurasia from about 300 kya to about 30 kya. Neanderthals had large brains and elongated skulls with very large faces. They also were characterized by very robust bodies.

Neocortex

part of the cerebral cortex and generally thought to be most closely associated with problem solving and behavioral flexibility. In mammals, the neocortex covers virtually the entire surface of the forebrain.

Neocortex Ration

the size of the neocortex in relation to the rest of the brain.

Neutral Theory

the theory that describes genetic change that is caused only by mutation and drift.

Nocturnal

describes an animal active only during the nighttime. See also diurnal.

Observational Learning

a form of learning in which animals observe the behavior of other individuals and thereby learn to perform a new behavior.

Oceania

a region of the world that includes Polynesia, Melanesia, and Micronesia.

Olfaction (olfactory, adj.)

the sense of smell.

Orbits

the bony sockets of the eye.

Organelle

a portion of the cell that is enclosed in a membrane and has a specific function; examples are mitochondria and the nucleus.

Out Of Africa Model

holds that modern humans evolved in Africa sometime during the last 100,000 years and spread out from that continent to replace other existing hominids in Europe and Asia.

Outgroups

taxonomic groups that are related to a group of interest and can be used to determine which traits are ancestral was which are derived.

Pairwise

describes an interaction that involves two individuals.

Paleontologist

a scientist who studies fossilized remains of plant and animals species.

Pangaea

the massive single continent that contained all of the Earth's dry land until about 180 mya.

Parent-offspring Conflict

conflicts that arise between parents and offspring over how much the parents will invest in their offspring. These conflicts stem from the opposing genetic interests of parents and offspring.

Perisylvain Region

the region of the brain close to the sylvian fissure. Language production is strongly affected by damage to the perisylvian region of the left hemisphere in humans.

Phenotypic Gambit

the research tactic used by evolutionary biologists to generate hypotheses about behavior by assuming that behavior is adaptive. Purposely neglected in these hypotheses are constraints, correlated character, genetic drift, and other factors that lead to maladaptive outcomes.

Phonemes

the basic unit speech perception. Phonemes are the smallest bits of sound that we recognize as meaningful elements of language. People who speak different languages recognize slightly different sets of phonemes.

Porphyria Variegate

a genetic disease caused by a dominant gene in which carriers of the gene develop a severe reaction to certain anesthetics.

Potassium-Argon Dating

a method used for dating volcanic rocks.

Premenstrual Syndrome (PMS)

a collection of symptoms that occur in some women just before menstruation begins, including irritability, moodiness, fatigue, and headaches.

Premolars

the teeth that lie between the canines and molars.

Primary Structure

the sequence of amino acids that make up a protein.

Progesterone

an ovarian steroid that plays an important role in preparing the fetus to sustain a pregnancy. Progesterone promotes the development of glands in the endometrium that secrete glycogen and important enzymes into the uterus.

Prokaryotes

organisms without a cell nucleus or separate chromosomes. Bacteria are prokaryotes.

Proteins

large molecules that consist of a long chain of amino acids. Many proteins are enzyme catalysis, while other proteins perform structural functions.

Radioactive Decay

spontaneous change from one isotope of an element to another isotope of the same element or to an entirely different element. Radioactive decay occurs at a constant rate that can be measured precisely in the laboratory.

Radiocarbon Dating Radiometric Methods

dating methods taking advantage of the fact that isotopes of certain elements change spontaneously from one isotope to anther at a constant rate.

Recessive

describes an allele that is expressed in the phenotype only when it is in the homozygous state.

Recombination

the creation of novel genotypes as a result of the random segregation of chromosomes and of crossing over.

Reinforcement

the process in which selection acts against the likelihood of hybrids occurring between members of two phenotypically distinctive populations leading to the evolution of mechanisms that prevent interbreeding.

Reproductive Isolation

the relationship between two populations when there is no gene flow between them.

Ribonucleic Acid (RNA)

a long molecule that plays several different important roles in protein synthesis. RNA differs from DNA in having a slightly different chemical backbone and the base uracil is substituted for thymine.

Ribosomes

small organelles composed of protein and nucleic acid that temporarily hold the messenger RNA and transfer RNAs together during proteins synthesis.

Selfish Behavior

behavior that increases the fitness of the donor and decreases the fitness of the recipient.

Sexual Dimorphism

differences between sexually mature males and females in body size or morphology.

Sexual Selection

a form of natural selection that results from differential mating success in one gender. In mammals, sexual selection usually occurs in males and may be due to male-male competition.

Sickle-Cell Anemia

a severe form of anemia that afflicts people who are homozygous for the sickle-cell gene.

Social Intelligence Hypothesis
the hypothesis that the relatively sophisticated cognitive abilities of higher primates are the outcome of selective pressures that favored intelligence as a means to gain advantages in social groups.

Species (sing. And pl.)
a group of organisms classified together at the lowest level of the taxonomic hierarchy. Biologists disagree about how to define a species.

Spinal Cord
the massive bundle of nerves that connects the brain to most of the rest of the body. The spinal cord passes down the vertebral canal.

Spite (spiteful, adj.)
behavior that is costly to the actor and the recipient.

Strategy
a complex of behaviors deployed in a specific functional context, such as mating, parenting, or foraging.

Syntax
the grammatical rules that allow people to assign meaning to strings of words.

Systematics
a branch of biology that is concerned with the procedures for construction phylogenies. See also taxonomy.

Taxonomy
a branch of biology that is concerned with the use of phylogenies for naming and classifying organisms. See also systematics.

Terrestrial

predominantly active on the ground.

Territories

fixed areas occupied by animals who defend the boundaries against intrusion by other individuals or groups of the same species.

Thymine

one of the four bases of the DNA molecule. The complementary base of thymine is adenine.

Traits

characteristics of organisms

Transfer RNA (tRNA)

a form of RNA that facilitates protein synthesis by first binding to amino acids in the cytoplasm and then binding the appropriate site on the mRNA molecule. There is at least one distinct form of RNA for each amino acid.

tRNA

see transfer RNA.

Unlinked

refers to genes on different chromosomes. See also linkage.

Upper Paleolithic

the period from about 40 kya to about 10 kya in Europe, North Africa, and parts of Asia. The tool kits form this period are dominated by blades.

Uracil

one of the four bases of the RNA molecule. It corresponds to the base thymine in DNA; as with DNA, its complementary base is adenine.

Variant

the particular form of a trait. For example, blue eyes, brown eyes, and gray eyes are variants of the trait eye color.

Variation Among Groups

differences in the average phenotype or genotype between groups.

Variation Between Groups

differences in the average phenotype or genotype between groups.

Vertebrae

the bones of the spinal column.

Vertical Clinging and Leaping

a form of locomotion in which the animal clings to vertical supports and moves by leaping from vertical support to another.

Vestibular System

a system to tubes that are embedded in the inner ear inside the cranium) and part of the system animals use to maintain balance.

Vitamins

organic substances necessary in small quantities for normal metabolism.

Viviparity

giving birth to live young.

Weak Garden of Eden Hypothesis

the view that the genes that give rise to modern human morphology arose in Africa between 100 kya and 200 kya. The people carrying these genes then spread throughout Africa and differentiated into a number of genetically variable populations. Then, around 50 kya, one of these African populations spread to the rest of the world, replacing the resident hominids.

Zygomatic Arches

the cheekbones.

Zygote

the cell formed by the union of an egg and sperm.

File Image No. 56

http://www.google.com/imgres?imgurl=http://1.bp.blogspot.com/-

NEHe35lyY9M/UK0G0FN7giI/AAAAAAAAGrw/wvbmefFxy7s/s400/blondeme
lanesians.jpg&imgrefurl=http://www.nancytoussaint.org/2012_11_01_archive.ht
ml&h=334&w=500&sz=19&tbnid=lJauwup5_yguBM:&tbnh=77&tbnw=116&zo
om=1&usg=__VHeRylGtt12PLi5KSm7C7fqkgw=&docid=Equn6Ifmyp_5OM&s
a=X&ei=wKP5UfzkJ5CyigKvxoH4CQ&ved=0CFcQ9QEwBw&dur=76

File Image No. 57

http://www.google.com/imgres?imgurl=http://1.bp.blogspot.com/ -

NEHe35lyY9M/UK0G0FN7giI/AAAAAAAAGrw/wvbmefFxy7s/s400/blondeme
lanesians.jpg&imgrefurl=http://www.nancytoussaint.org/2012_11_01_archive.ht
ml&h=334&w=500&sz=19&tbnid=lJauwup5_yguBM:&tbnh=77&tbnw=116&zo
om=1&usg=__VHeRylGtt12PLi5KSm7C7fqkgw=&docid=Equn6Ifmyp_5OM&s
a=X&ei=wKP5Ufzk J5CyigKvxoH4CQ&ved=0CFcQ9QEwBw&dur=76

File Image No. 54

http://www.google.com/imgres?imgurl=http://blogs.discovermagazine.com/gnx
p/files/2010/08/Populations_first_wawe_migr299x300.png&imgrefurl=http://bl
ogs.discovermagazine.com/gnxp /2010/08/blondes-of-the
blackislands/&h=300&w=299&sz=169&tbnid=zKqBedaKMcGrmM:&tbnh=90&
tbnw=90&zoom=1&usg=__peojo2aX106pPFNWaj5fIGgY9Eo=&docid=dCnoD
ppHRDtDWM&sa=X&ei=gl_9UZ76MouOiglLB
goHoBw&ved=0CFEQ9QEwBg&dur=524

File Image No. 57

http://www.google.com/imgres?imgurl=http://blogs.discovermagazine.com/gnx
p/files/2010/08/Populations_first_wawe_migr299x300.png&imgrefurl=http://bl
ogs.discovermagazine.com/gnxp/2010/08/blondes-of
theblackislands/&h=300&w=299&sz=169&tbnid=zKqBedaKMcGrmM:&tb
nh=90&tbnw=90&zoom=1&usg=__peojo2aX106pPFNWaj5fIGgY9
Eo=&docid=dCnoDppHRDtDWM&sa=X&ei=gl_9UZ76MouOigLB
goHoBw&ved=0CFEQ9QEwBg&dur=524#imgdii=zKqBedaKMcGr
mM%3A%3BIBwngDEVX5loEM%3BzKqBedaKMcGrmM%3A

File Image No. 42

*http://www.google.com/imgres?imgurl=http://resources1.news.com.au/images/20
10/12/23/1225975/325457-*

denisovans2.jpg&imgrefurl=http://www.theaustralian.com.au/news/healthscienc
e/meet-the-denisovans-indigenous-australiassiberian-kin/story-e6frg8y6-
1225975312349&h=366&w=650&sz=70&tbnid=A09DlAuSvxL1W
M:&tbnh=67&tbnw=119&zoom=1&usg=__fGaSEOvvThllkTXASeF
kfgRQFT4=&docid=5IHnqH9eL8xSuM&sa=X&ei=6GH9Ua20M4f
XigLo9IGwBg&sqi=2&ved=0CEwQ9QEwBA&dur=1

File Image No. 75

http://www.google.com/imgres?imgurl=http://static.ddmcdn.com /gif/designer-

childrenbasics.gif&imgrefurl=http://science.howstuffworks.com/life/geneti
c/designer-
children1.htm&h=430&w=372&sz=31&tbnid=sb_izD45jpUsM:&tbnh=81&tbn
w=70&zoom=1&usg=__da6ozvLI5zkXM ABJ3K3ew6AFF-
Y=&docid=RB839WOaIiHOM&sa=X&ei=cmX9Uc3YAYbCigKe9oAQ&ved=0
CG IQ9QEwEQ&dur=153

File Image No. 76

http://www.google.com/imgres?imgurl=http://www.globalchange.

umich.edu/gctext/Inquiries/Inquiries_by_Unit/Unit_5_files/image015.jpg&imgr
efurl=http://www.globalchange.umich.edu/gctext/Inquiries/Inquiries_by_Unit/
Unit_5.htm&h=414&w=588&sz=23&tbnid=MSRZuBBv_9uGKM:&tbnh=82&tb
nw=116&zoom=1&usg=__Q73V52JTmA10yqCbN5XHn78ks=&docid=gjMpy3
aVFjHKXM&sa=X&ei=tWj9UcHGBqrIiwL4l4CIBg&ved=0CDIQ9QEwAg&dur
=385

File Image No. 73

http://www.google.com/imgres?imgurl=http://www.racialcompact.com/Phyloge
neticTree.jpg&imgrefurl=http://www.racialcompact.com/Race_%2520Realityan
d%2520Denial.html&h=614&w=589&sz=35&tbnid=XySxR0l8AEmmVM:&tbnh
=88&tbnw=84&zoom=1&usg=__JpmWlcBRIAKQBx8vJKVl5ulYdXY=&docid=
KX5F6XRcWwVCzM&sa=X&ei=u2r9Uba5DIGCiwK2koCQCw&ved=0CEwQ9
QEw Cg&dur=1

File Image No. 40

http://www.google.com/imgres?imgurl=http://www.encognitive.com/files/images
/human-evolution-timeline-chart-

treetheory.jpg&imgrefurl=http://www.encognitive.com/node/10796&h
=323&w=640&sz=100&tbnid=sCVWkyAviTtOoM:&tbnh=59&tbnw
=116&zoom=1&usg=__5xvVPt1C__dqsNBEQ7STmkfGWMc=&doc
id=MGFb#imgdii=sCVWkyAviTtOoM%3A%3BFwBc0nhcDJIHM%3BsCVWky
AviTtOoM%3AbbyUNHLi6M&sa=X&ei=fRjrUavN
K8bKiwKUuYC4CA&ved=0CC8Q9QEwAQ&dur=489#imgdii=sCV
WkyAviTtOoM%3A%3BFwBc0nhcDJIHM%3BsCVWkyAviTtOoM%3A

File Image No. 12

http://biology.unm.edu/ccouncil/Biology_203/Summaries/Phylog eny.htm

File Image No. 15

http://www.google.com/imgres?imgurl=http://upload.wikimedia.o
rg/wikipedia/commons/e/ed/CharlesDarwin31.jpg&imgrefurl=http://commons.
wikimedia.org/wiki/File:Charle s-
Darwin31.jpg&h=936&w=844&sz=330&tbnid=BRXR0CqUkTZuZM:&tbn
h=186&tbnw=167&zoom=1&usg=__oCB1DvMscWXGKigFc3bo9ouXQ=&doci
d=nFBx1igQfT6wM&sa=X&ei=qnD9UY24N8XHigLa04CwDg&ved=0CC0
Q9QEwAA&dur=46

http://photobucket.com/images/charles%20darwin?page=1

File Image No. 17

http://www.google.com/imgres?imgurl=http://www.rarelibrary.com/authorimag
es/darwin.jpg&imgrefurl=http://rarelibrary.com/biography/Charles%2BDarwin.
html&h=400&w=294&sz=39&tbnid=psfNhUrSghuiiM:&tbnh=95&tbnw=70&zo
om=1&usg=__LTgeC58vW2aEqFFVmqjZsJ4DNHM=&docid=9jILLDD2kMlcI
M&sa=X&ei=m3H9Ub6YFI3riQKUwoCoDw&ved=0CE4Q9QEwCg&dur=1195

File Image No. 17

http://www.sciencekids.co.nz/pictures/scientists/charlesdarwin.ht ml

File Image No. (A)

http://www.google.com/imgres?imgurl=http://daily.swarthmore.edu/static/uploads/by_date/2009/02/19/evolution.jpg&imgrefurl=http://www.atheismresource.com/evolution&h=480&w=640&sz=48&tbnid=M3dM_SXxhVjeIM:&tbnh=101&tbnw=134&zoom=1&us=__JfpMG4P0-F2ptbq0MwVN_bbkzWc=&docid=MlRHjFE18wHW7M&sa=X&ei=RHL9UczCMOazigKt44DICg&ved=0CC8Q9QEwAA&dur=0http://www.atheismresource.com/evolution

https://www.google.com/search?site=&source=hp&q=charles+darwin+images&oq=Darwin+Images&gs_l=hp.1.6.0l3j0i22i30l7.5194.13805.0.19313.15.14.1.0.0.0.266.1807.0j13j1.14.0....0...1c.1.19.hp.mprIkhKzcg#bav=on.2,or.r_qf.&ei=m3H9Ub6YFI3riQKUwoCoDw&fp=f3f8a15b7409e0a4&psj=1&q=evolution+images&revid=1466208725&sa=X&ved=0CJQBENUCKAE

File Image No. 67

http://www.asianplasticsurgeryguide.com/beforeafter/rhinoplastydorsaltip.html

File Image No. 37

http://darwiniana.org/hominid.htm

File Image No. 38

http://darwiniana.org/hominid.htm

File Image No. 61

https://www.google.com/search?q=khoisan+people+look+asian&tbm=isch&tbo=u&source=univ&sa=X&ei=zu7iUf_vEYeGiQKJpYDQBQ&ved=0CDYQsAQ&biw=1149&bih=697#facrc=_&imgdii=_&imgrc=0NF2EWKuwRAEfM%3A%3BCINNi5qwmZxHJM%3Bhttp%253A%252F%252Fi45.tinypic.com%252F2ed9n4j.jpg%3Bhttp%253A%252F%252Fwww.nairaland.com%252F347410%252Fpicturethread-1-khoisan-people%3B209%3B261

File Image No. 62

https://www.google.com/search?q=khoisan+people+look+asian&tbm=isch&tbo=u&source=univ&sa=X&ei=zu7iUf_vEYeGiQKJpYDQBQ&ved=0CDYQsAQ&biw=1149&bih=697#facrc=_&imgdii=_&imgrc=LfMH3L8Fz7VbnM%3A%3BCINNi5qwmZxHJM%3Bhttp%253A%252F%252Fi45.tinypic.com%252F2qjhsfp.jpg%3Bhttp%253A%252F%252Fwww.nairaland.com%252F347410%252Fpicturethread-1-khoisan-people%3B192%3B254

File Image No. 66

https://www.google.com/search?q=khoisan+people+look+asian&tbm=isch&tbo=u&source=univ&sa=X&ei=zu7iUf_vEYeGiQKJpYDQBQ&ved=0CDYQsAQ&biw=1149&bih=697#facrc=_&imgdii=_&imgrc=QLWOxKPIIStNkM%3A%3BY1pZHmdlEiNOAM%3Bhttp%25 3A%252F%252Fhtl-wireless.com%252Fbroadnose.jpg%3Bhttp%253A%252F%252Fhtlwireless.com%252Fcolor.html%3B198%3B236

File Image No. 51

https://www.google.com/search?q=khoisan+people+look+asian&tbm=isch&
tbo=u&source=univ&sa=X&ei=zu7iUf_vEYeGiQKJpYDQBQ&ved=0CDYQ
sAQ&biw=1149&bih=697#facrc=_&imgdii=_&imgrc=Is9tiu6I5OZEKM%3A
%3BBlO3clOBMALToM%3Bhttp%253A%252F%252Frealhistoryww.com%2
52Fworld_history%252Fancient%252FImages_Thrace%252FBlonde_black%
252FBlue_eyes_4.jpg%3Bhttp%253A%252F%252Frealhistoryww.com%252F
world_history%252Fancient%252Fcro_magnon_Homo_sapien.htm%3B423
%3B312

File Image No. 31

https://www.google.com/search?q=khoisan+people+look+asian&tbm=isch&
tbo=u&source=univ&sa=X&ei=zu7iUf_vEYeGiQKJpYDQBQ&ved=0CDYQ
sAQ&biw=1149&bih=697#facrc=_&imgdii=_&imgrc=aL1yFmP9EdVpgM%
3A%3BBlO3clOBMALToM%3Bhttp%253A%252F%252Frealhistoryww.com
%252Fworld_history%252Fancient%252Fimages_eman%252FNeanderthal_c
hild.jpg%3Bhttp%253A%252F%252Frealhistoryww.com%252Fworld_history
%252Fancie nt%252Fcro_magnon_Homo_sapien.htm%3B307%3B424

File Image No. 59

https://www.google.com/search?q=khoisan+people+look+asian&tbm=isch&
tbo=u&source=univ&sa=X&ei=zu7iUf_vEYeGiQKJpYDQBQ&ved=0CDYQ
sAQ&biw=1149&bih=697#facrc=_&imgdii=_&imgrc=J3J7qHI9Wex2CM%3
A%3B9gWMLF0YouCE0M%3Bhttp%253A%252F%252Fi.imgur.com%252F
LjUzk.jpg%3Bhttp%253A%252F%252Fwww.theapricity.com%252Fforum%2
52Farchive%252Find ex.php%252Ft-56206.html%3B288%3B384

File Image No. 60

https://www.google.com/search?q=khoisan+people+look+asian&tbm=isch&tbo=u&source=univ&sa=X&ei=zu7iUf_vEYeGiQKJpYDQBQ&ved=0CDYQsAQ&biw=1149&bih=697#facrc=_&imgdii=_&imgrc=vlnlc7huM7RqM%3A%3BCINNi5qwmZxHJM%3B http%253A%252F%252Fi47.tinypic.com%252F2qwie8h.jpg%3Bhttp%253A%252F%252Fwww.nairaland.com%252F347410%252Fpicture-threadkhoisanpeople%3B192%3B239

File Image No. 77

http://users.rcn.com/jkimball.ma.ultranet/BiologyPages/D/DNAR eplication.html

File Image No. 69

http://upload.wikimedia.org/wikipedia/commons/b/b2/Karyotype .png

File Image No. 71

http://en.wikipedia.org/wiki/File:Mitochondrial_DNA_en.svg

File Image No. 35

http://en.wikipedia.org/wiki/Cro-Magnon

File Image No. 36

http://en.wikipedia.org/wiki/Cro-Magnon

File Image No. 31

http://realhistoryww.com/world_history/ancient/cro_magnon_H omo_sapien.htm

File Image No. 43

http://realhistoryww.com/world_history/ancient/cro_magnon_H omo_sapien.htm

File Image No. 60

http://realhistoryww.com/world_history/ancient/cro_magnon_H omo_sapien.htm

File Image No. 63

http://realhistoryww.com/world_history/ancient/cro_magnon_H omo_sapien.htm

File Image No. 61

http://realhistoryww.com/world_history/ancient/cro_magnon_H omo_sapien.htm

File Image No. 64

http://realhistoryww.com/world_history/ancient/cro_magnon_H omo_sapien.htm

File Image No. 62

http://realhistoryww.com/world_history/ancient/cro_magnon_H omo_sapien.htm

File Image No. 59

http://realhistoryww.com/world_history/ancient/cro_magnon_H omo_sapien.htm

File Image No. 46

http://realhistoryww.com/world_history/ancient/cro_magnon_H omo_sapien.htm

File Image No. 47

http://realhistoryww.com/world_history/ancient/cro_magnon_H omo_sapien.htm

File Image No. 48

http://realhistoryww.com/world_history/ancient/cro_magnon_H omo_sapien.htm

File Image No. 49

http://realhistoryww.com/world_history/ancient/cro_magnon_H omo_sapien.htm

File Image No. 51

File Image No. 62

http://realhistoryww.com/world_history/ancient/cro_magnon_H omo_sapien.htm

File Image No. 65

http://realhistoryww.com/world_history/ancient/cro_magnon_H omo_sapien.htm

File Image No. 21

http://www.google.com/imgres?imgurl=http://4.bp.blogspot.com/_PRWlq_BwP2E/S6a
c2bstCI/AAAAAAAABsE/LKBj6Cboqs4/s200/neanderthal.jpg&imgrefurl=http://esta
mos-vivo.blogspot.com/2010/05/embraceyour-inner-
neanderthal.html&h=381&w=300&sz=22&tbnid=TM0i9sO5mVHAM:&tbnh=97&tbn
w=76&prev=/search%3Fq%3DCroMagnon%2BMan%2BPics%26tbm%3Disch%26tbo
%3Du&zoom=1&q=CroMagnon+Man+Pics&usg=__6QPcR9ccm6oH6pf4qAPYX_i69
7M=&docid=HiFVEpRGh3ZZAM&hl=en&sa=X&ei=1I79UcapAWXiQKH5YCADg&
ved=0CGgQ9QEwEw&dur=505

File Image No. 33

http://www.google.com/imgres?imgurl=http://mariecachet.files.wordpress.com/2012/0
9/neandertal6.jpg&imgrefurl=http://atala.fr/2012/09/24/2neanderthalen/&h=328&w=6
00&sz=18&tbnid=hVtFlnMBlSon5M:&tbnh=66&tbnw=120&prev=/search%3Fq%3Dn
eanderthal%2Bcro%2Bmagnon%2Btimeline%26tbm%3Disch%26tbo%3Du&zoom=1&
q=neander thal+cro+magnon+timeline&usg=__QSxkvf6iG3mQofO9VRNzVrP
Qhs0=&docid=L40OzVnOgX748M&hl=en&sa=X&ei=25D9UaPDGaStigKRvoGwCw
&ved=0CFIQ9QEwBQ&dur=106 HOMOSAPIEN – The Emergence and Evolution of
the Human species

File Image No. 77

http://www.google.com/search?q=p+gene+albinism&hl=en&sa=X&tbm=isch&tbo=u
&source=univ&ei=gMMrUYINHniALHmIGIBg&ved=0CGUQsAQ&biw=1149&bih=
726#facrc=_& imgdii=_&imgrc=UGJJEREIZXBsCM%3A%3BbGZDzEZRGFJwq
M%3Bhttp%253A%252F%252Frealhistoryww.com%252Fworld_history%252Fancient

Image Credit Citations

File Image No. 78

http://www.google.com/search?q=p+gene+albinism&hl=en&sa=X&tbm=isch&tbo=u
&source=univ&ei=gMMrUYINHniALHmIGIBg&ved=0CGUQsAQ&biw=1149&bih=
726#facrc=_& imgdii=_&imgrc=V_rBbzOppsIC_M%3A%3Bf2qPwl3I7XJBEM%3
Bhttp%253A%252F%252Fdrugline.org%252Fimg%252Fail%252F131_132_3.jpg%3Bht
tp%253A%252F%252Fdrugline.org%252Fail%252Fpathography%252F132%252F%3B3
80%3B411

File Image No. 79

http://www.nutralegacy.com/blog/general-healthcare/6-factsabout-albinism/
HOMOSAPIEN – The Emergence and Evolution of the Human species

File Image No. 78

http://www.google.com/search?q=p+gene+albinism&hl=en&sa=X&tbm=isch&tbo=u
&source=univ&ei=gMMrUYINHniALHmIGIBg&ved=0CGUQsAQ&biw=1149&bih=
726#facrc=_& imgdii=_&imgrc=V_rBbzOppsIC_M%3A%3Bf2qPwl3I7XJBEM%3
Bhttp%253A%252F%252Fdrugline.org%252Fimg%252Fail%252F131_132_3.jpg%3Bht
tp%253A%252F%252Fdrugline.org%252Fail%252Fpathography%252F132%252F%3B3
80%3B411

File Image No. 85

http://www.google.com/search?q=p+gene+albinism&hl=en&sa=X&tbm=isch&tbo=u
&source=univ&ei=gMMrUYINHniALHmIGIBg&ved=0CGUQsAQ&biw=1149&bih=
726#facrc=_& imgdii=_&imgrc=PGEK6ODh1hoBcM%3A%3BqYnqHZ1zegml7M
%3Bhttp%253A%252F%252Fwww.dermaamin.com%252Fsite%252Fimages%252Fclini
calpic%252Fa%252Falbinism%252Falbinism8.jpg%3Bhttp%253A%252F%252Fwww.de
rmaamin.com%252Fsite%252Fatlas-ofdermatology%252F1-a%252F86-albinism-
.html%3B500%3B750

File Image No. 86

http://www.google.com/search?q=p+gene+albinism&hl=en&sa=X&tbm=isch&tbo=u
&source=univ&ei=gMMrUYINHniALHmIGIBg&ved=0CGUQsAQ&biw=1149&bih=
726#facrc=_&imgdii=_&imgrc=KBLrkNLuG5YLM%3A%3BkBtFpG1N7T9UzM%3Bh
ttp%253A%252F%252Frealhistoryww.com%252Fworld_history%252Fancient%252FM
isc%252FC ommon%252FIndia%252FIndian_Albinos%252FV_P_Sekar_s.jpg
%3Bhttp%253A%252F%252Fblacktoday.wordpress.com%252F2013%252F07%252F03
%252Fwhite-history%252F%3B240%3B304

File Image No. 79

http://www.google.com/search?q=p+gene+albinism&hl=en&sa=X&tbm=isch&tbo=u
&source=univ&ei=gMMrUYINHniALHmIGIBg&ved=0CGUQsAQ&biw=1149&bih=
726#facrc=_& imgdii=_&imgrc=YOWQciIYR5aUGM%3A%3Bp2snMnhCmrNjqM
%3Bhttp%253A%252F%252Fwww.carlin.co.nz%252FPhotos%252FSI_Albino_Baby.jp
g%3Bhttp%253A%252F%252Fwww.medbullets. com%252Fstep2-3-
dermatology%252F20075%252Falbinism%3B400%3B600

File Image No. 01

http://0.tqn.com/d/dinosaurs/1/0/i/J/-/-/leedsichthysDB.jpg

File Image No. 02

http://0.tqn.com/d/dinosaurs/1/0/P/G/-/-/crassigyrinusDB.jpg

File Image No. 03

http://0.tqn.com/d/dinosaurs/1/0/Y/I/-/-/cacopsDB.jpg

File Image No. 04

http://dinosaurs.about.com/od/otherprehistoriclife/tp/VertebrateAnimal-Evolution.htm

File Image No. 05

http://0.tqn.com/d/dinosaurs/1/0/N/3/-/-/grippia.jpg HOMOSAPIEN – The Emergence and Evolution of the Human species

File Image No. 06

http://0.tqn.com/d/dinosaurs/1/0/r/1/-/-/thalassodromeus.jpg

File Image No. 07

http://0.tqn.com/d/dinosaurs/1/0/y/F/-/-/gastornisWC.jpg

File Image No. 09

http://0.tqn.com/d/dinosaurs/1/0/C/6/-/-/megazostrodon.jpg

File Image No. 10

http://0.tqn.com/d/dinosaurs/1/0/_/K/-/-/sarkastodonDB.jpg

File Image No. 11

http://0.tqn.com/d/dinosaurs/1/0/u/O/-/-/hadropithecusWC.jpg

References

[1][1] Smithsonian Intimate Guide to Human Origins - Carl Zimmerman

[1][2] Evolution – A Very Short Introduction – Brian & Deborah Charlesworth

[1][3] Evolution – Cro-Magnon – Brian Fagan

[1[[4]Human Evolution – A Very Short Introduction - Wood, Bernard

[1][5] http://genographic.nationalgeographic.com

[1][6] The Vertebrate Animal Evolution - From Fish to Primates - Bob Strauss, About.com Guide - http://dinosaurs.about.com/od/otherprehistoriclife/tp/VertebrateAnimal-Evolution.htm

[1][7] Neanderthal Genome– Ch 11http://www.sciencemag.org/site/special/neandertal/feature/index.htm l

[1][8] (wikipedia - Phylogenetics – Reconstructing the tree of life) http://en.wikipedia.org/wiki/Phylogenetics

[1][9] Ch 12- DNA hybridization http://en.wikipedia.org/wiki/DNA-DNA_hybridization

[1][10] Genetic variation between humans - Wikipedia: Reference desk/Archives/Science/2012 July 7 http://en.wikipedia.org/wiki/Wikipedia:Reference_desk/Archives/Scien ce/2012_July_7

[1][11] Ch 12Neanderthal and Modern Human Interbreeding - The first genetic code of Neanderthal reveals interbreeding - http://www.nhm.ac.uk/about-us/news/2010/may/first-genetic-code-ofneanderthal-reveals-interbreeding66724.html - 31 August 2012

[1][12] Ch 13 National Geographic's Genographic Project
https://genographic.nationalgeographic.com/about/

[1][13] Ch 12 DNA and Genes- The structure of DNA and GenesDeoxyribonucleic
Acid (DNA) http://www.genome.gov

[1][14] Ch 12 DNA and Genes-New DNA analysis shows ancient humans interbred
with Denisovanshttp://www.nature.com/news/new-dna-analysisshows-ancient-
humans-interbred-with-denisovans-1.11331 - An article from Scientific American

[1][15] Ch 12 -12.3 Denisovans, indigenous Australia's Siberian kin (Meet the
Denisovans, indigenous Australia's Siberian kin)
http://www.theaustralian.com.au/news/health-science/meet-thedenisovans-
indigenous-australias-siberian-kin/story-e6frg8y6- 1225975312349

[1][16] Introduction The Keys to Discovery
https://genographic.nationalgegraphical.com

[1][17] Ch 13 Phylogeny
http://biology.unm.edu/ccouncil/Biology_203/Summaries/Phylogeny.ht m

[1][18] CH12 - Denisovans, indigenous Australia's Siberian kin
http://www.theaustralian.com.au/subscribe/digitalpass/2/

[1][19] CH 12 - Human Evolution Timeline http://www.proof-of-
evolution.com/human-evolution-timeline.html

[1][20] Preface- Homosapien
http://www.urbandictionary.com/define.php?term=homosapien

www.ingramcontent.com/pod-product-compliance
Lightning Source LLC
Chambersburg PA
CBHW071318210326
41597CB00015B/1264